APPLICATIONS
of CHITIN
and CHITOSAN

APPLICATIONS
of CHITIN
and CHITOSAN

Edited by

MATTHEUS F. A. GOOSEN, Ph.D

CRC PRESS

Boca Raton   London   New York   Washington, D.C.

Library of Congress Cataloging-in-Publication Data

Main entry under title:
  Applications of Chitin and Chitosan

Visit the CRC Web site at www.crcpress.com

© 1997 by CRC Press LLC

No claim to original U.S. Government works
International Standard Book Number 1-56676-449-1
Library of Congress Card Number 96-61229

Contents

Preface   xi

PART I: OVERVIEW

1. **Applications and Properties of Chitosan**   3
   Q. LI–*Queen's University*
   E. T. DUNN–*Queen's Univeristy*
   E. W. GRANDMAISON–*Queen's University*
   M. F. A. GOOSEN–*Queen's University*

   Introduction 3
   Physicochemical Properties of Chitosan 5
   Applications of Chitosan 9
   Concluding Remarks 21
   References...21

2. **Applications of Chitin and Chitosan in the**   31
   Ecological and Environmental Fields
   SHIGEHIRO HIRANO–*Tottori Univeristy*

   Introduction 31
   Molecular Characteristics of Chitin and Chitosan 31
   Current Commercial Uses of Chitin and Chitosan 32
   Environmentally and Ecologically
        Friendly Applications................................34
   Conclusions 48

Acknowledgements 49
References 50

PART II: STRUCTURE AND PROPERTIES

3. **Chitin Structure and Activity of**   57
   Chitin-Specific Enzymes
   MARIA L. BADE – *Massachusetts Institute of Technology*

   Conclusions 73
   Acknowledgements 75
   References...75

4. **β-Chitin and Reactivity Characteristics**   79
   KEISUKE KURITA – *Seikei University*

   Introduction 79
   Isolation of β-Chitin and Some Properties...................80
   Reactions of β-Chitin 80
   Reaction of Chitosan Derived from β-Chitin 84
   Conclusion 86
   References...86

5. **Characterization and Solution Properties of**   89
   Chitosan and Chitosan Derivatives
   M. RINAUDO – *Joseph Fourier University*
   M. MILAS – *Joseph Fourier University*
   J. DESBRIÈRES – *Joseph Fourier University*

   Introduction 89
   Experimental 89
   Results and Discussion 90
   Conclusion 100
   References...101

6. **Soluble Precursors for Efficient Chemical**   103
   Modifications of Chitin and Chitosan
   KEISUKE KURITA – *Seikei University*

   Introduction 103
   Water-Soluble Chitin 104
   Tosyl- and Iodo-Chitins 106
   N-Phthaloyl-Chitosan 108
   Conclusion 111
   References...111

PART III: FOOD AND AGRICULTURE

7. **Chitosans as Dietary Food Additives**   **115**
   RICCARDO A. A. MUZZARELLI–*University of Ancona*
   MASSIMO DE VINCENZI–*Instituto Superiore di Sanità*

   Hypocholesterolemic Action of Animals 115
   Hypocholesterolemic Effect of Chitosan in Humans 120
   Perspectives for the Treatment of Celiac Disease 120
   Hypouricemic Effect of Chitin and Chitosan 123
   Conclusive Remarks 124
   Acknowledgement 125
   References 125

8. **Enhancing Food Production with Chitosan**   **129**
   Seed-Coating Technology
   DONALD FREEPONS–*Kennewick, WA*

   Introduction 129
   History of Phytohormones 131
   Chitosan Biochemistry 132
   Chitosan Benefits 133
   Mode of Action 135
   Summary 138
   References 138

9. **Inhibition of Molting in Chewing Insect Pests**   **141**
   MARIA L. BADE–*Massachusetts Institute of Technology*

   Conclusions 153
   Acknowledgements 153
   References 153

10. **Properties of Insect Chitin Synthase: Effects**   **155**
    of Inhibitors, γ-S-GTP, and Compounds
    Influencing Membrane Lipids
    MICHAEL LONDERSHAUSEN–*Institute for Parasitology*
    ANDREAS TURBERG–*Institute for Parasitology*
    MONIKA LUDWIG–*Heinrich-Heine University*
    BÄRBEL HIRSCH–*Heinrich-Heine University*
    MARGARETHE SPINDLER-BARTH–*Heinrich-Heine University*

    Introduction 155
    Biological Activity of Chitin Synthesis Inhibitors 156
    Effects of Inhibitors, γ-S-GTP, and Compounds
       Interacting with Membrane Lipids on Chitin

Biosynthesis in an Insect Cell Line from
Chironomus tentans 160
Summary 166
Conclusion 167
Abbreviations.................................... 167
Acknowledgements 167
References....................................... 168

11. New Applications of Chitin and Its Derivatives   171
in Plant Protection
HENRYK STRUSZCZYK—*Institute of Chemical Fibres*
HENRYK POSPIESZNY—*Institute of Plant Protection*

Introduction 171
Antibacterial Activity of Chitin and Its Derivatives 172
Antivirus Activity of Chitosan 174
Antiviroid Activity of Chitosan 178
Antiphage Activity of Chitosan 178
Biostimulation of Plant Growth by Chitosan 180
Conclusions 183
References....................................... 184

12. Chitinolytic Enzymes in Selected Species of   185
Veterinary Importance
MICHAEL LONDERSHAUSEN—*Institute for Parasitology*
ANDREAS TURBERG—*Institute for Parasitology*

Introduction 185
Methods in Investigation of Chitinolytic Enzymes 185
Characterization of *Lucilia cuprina* Chitinase 187
Chitinase Inhibitors 190
Molecular Weight Pattern of Chitinases from
Different Veterinary Parasites....................... 193
Summary 198
Abbreviations.................................... 198
Acknowledgements 199
References....................................... 199

PART IV: MEDICINE AND BIOTECHNOLOGY

13. Applications of Chitin, Chitosan, and Their   205
Derivatives to Drug Carriers for Microparticulated
or Conjugated Drug Delivery Systems
HIRAKU ONISHI—*Hoshi University*
TSUNEJI NAGAI—*Hoshi University*
YOSHIHARU MACHIDA—*Hoshi University*

Introduction 205
Chitosan Microspheres 206
Chitosan-Drug Conjugates 212
Future Aspects 227
Acknowledgements 228
References 228

14. Chitosan-Alginate Affinity Microcapsules for   233
Isolation of Bovine Serum Albumin

MATTHEUS F. A. GOOSEN – *Sultan Qaboos University*
OSEI-WUSU ACHAW – *Queen's University*
EDWARD W. GRANDMAISON – *Queen's University*

Introduction 233
Preparation of Affinity Microcapsules 236
Adsorption Studies 238
Conclusions 250
Nomenclature 252
Acknowledgements 252
References 252

15. Structure and Isolation of Native Animal Chitins   255

MARIA L. BADE – *Massachusetts Institute of Technology*

Introduction 255
Chitin: Chemical Identity 255
Chitin: Structure 256
Chitin: Maintenance of Structure during
   Purification 260
Chitin: Optical Confirmation of Structure 269
Chitin: Chitin-Protein Interaction 271
Chitin: Parallel or Antiparallel 273
Chitin: Summary 274
Conclusions 274
Acknowledgements 275
References 275

PART V: TEXTILES AND POLYMERS

16. Fine Structure and Properties of Filaments   281
Prepared from Chitin Derivatives

G. W. URBANCZYK – *Technical University of Lodz*

Introduction 281
Experimental 281
Results 282

Conclusions 295
References..295

17. Graft Copolymers   297
KEISUKE KURITA—*Seikei University*

Introduction 297
Graft Copolymerization onto Chitin 298
Graft Copolymerization onto Chitosan 302
Conclusion 304
References..304

PART VI: WASTEWATER TREATMENT
18. Applications of Chitosan for the Elimination of   309
Organochlorine Xenobiotics from Wastewater
J. P. THOME—*University of Liège*
CH. JEUNIAUX—*University of Liège*
M. WELTROWSKI—*Textile Technology Centre*

Introduction 309
Use of Chitosan to Recover Organochlorinated
    Compounds 310
A Class of Priority Organochlorine
    Pollutants: The PCBs 311
Preliminary Studies of PCB Sorption on Chitin
    and Chitosan 313
PCB Sorption Ability of Chemically
    Modified Chitosan...............................314
PCB Sorption on Cross-Linked Chitosan as a
    Function of the Mesh Size and the PCB
    Chlorine Content 317
Evaluation of Reduced Cross-Linked Chitosan
    for Epuration of Large Volumes of
    PCB-Contaminated Stream Water 320
Pentachlorophenol (PCP) Adsorption on
    Chitosan Derivatives 322
Conclusions 326
References..327

Index   333

Preface

The main driving force behind the development of new applications for chitin and its derivative chitosan lies with the fact that these polysaccharides represent a renewable source of natural biodegradable polymers. Since chitin is the second most abundant natural polymer, academic and industrial scientists are faced with a great challenge to find new and practical applications for this material. This book is intended to fill this need. It will provide an examination of the state of the art. New applications, as well as potential products, will be examined.

The book is divided into six parts. For people who are new to the field, two special background/overview chapters, dealing with applications, structures, and properties of chitin and chitosan, are provided at the beginning of the book. This is followed by five specific application sections covering structure and properties, food and agriculture, medicine and biotechnology, textiles and polymers, and wastewater treatment.

The intended audience for this book includes industrial personnel involved in bioprocessing, as well as bioengineering students, specialists in the biomedical and biopharmaceutical industry, biochemists, food engineers, environmentalists, and microbiologists and biologists who wish to specialize in chitosan technology.

Our book will differ from competing volumes by dealing almost exclusively with applications. Previous books devoted less than 30% of their available space to commercial/medical uses.

OVERVIEW

CHAPTER 1

Applications and Properties of Chitosan

Q. Li, E. T. Dunn, E. W. Grandmaison, and M. F. A. Goosen*
Department of Chemical Engineering
Queen's University
Kingston, Ontario, Canada
K7L 3N6

INTRODUCTION

Chitosan is a polysaccharide obtained by deacetylating chitin, which is the major constituent of the exoskeleton of crustaceous water animals. This biopolymer was traditionally used in the Orient for the treatment of abrasions and in America for the healing of machete gashes [1]. A recent analysis of the varnish on one of Antonio Stradivarius's violins showed the presence of a chitinous material [2]. Chitosan was reportedly first discovered by Rouget in 1859 [3] when he boiled chitin in a concentrated potassium hydroxide solution. This resulted in the deacetylation of chitin. Fundamental research on chitosan did not start in earnest until about a century later. In 1934, two patents, one for producing chitosan from chitin and the other for making films and fibers from chitosan, were obtained by Rigby [4,5]. In the same year, the first X-ray pattern of a well-oriented fiber made from chitosan was published by Clark and Smith [6]. Since then, knowledge about chitosan has been greatly advanced by the work of such pioneers as Muzzarelli [3].

The main driving force in the development of new applications for

This chapter was originally published in the *Journal of Bioactive and Compatible Polymers*, October 1992.
*Present address, Sultan Qaboos University, College of Agriculture, P.O. Box 34, Al-Khod 123, Muscat, Sultanate of Oman.

3

chitosan lies in the fact that the polysaccharide is not only naturally abundant, but it is also nontoxic and biodegradable. Unlike oil and coal, chitosan is a naturally regenerating resource (e.g., crab and shrimp shells) that can be further enhanced by artificial culturing. It was reported that chitosan and chitin are contained in cell walls of fungi [7,8]. Chitin, however, is more widely distributed in nature than chitosan and can be found in mushrooms, yeasts, and the hard outer shells of insects and crustaceans. It was reported, for example, that about 50–80% of the organic compounds in the shells of crustacea and the cuticles of insects consists of chitin [3]. At present, most chitosan in practical and commercial use comes from the production of deacetylated chitin with the shells of crab, shrimp, and krill (the major waste by-product of the shellfish-processing industry) being the most available sources of chitosan [9,10].

One of the most useful properties of chitosan is for chelation. Chitosan can selectively bind desired materials such as cholesterol, fats, metal ions, proteins, and tumor cells. Chelation has been applied to areas of food preparation, health care, water improvement, and pharmaceutics. Chitosan has also shown affinity for proteins, such as wheat germ agglutinin [11] and trypsin [12]. Other properties that make chitosan very useful include inhibition of tumor cells [13], antifungal effects [14], acceleration of wound healing [15,16], stimulation of the immune system [17–19], and acceleration of plant germination [20].

Chitosan is a good cationic polymer for membrane formation. In early research it was shown that membranes formed from the polymer could be exploited for water clarification, filtration, fruit coating, surgical dressing, and controlled release. In 1978, for example, Hirano showed that N-acetyl chitosan membranes were ideal for controlled agrochemical release [21]. Later, he found that a semipermeable membrane with a molecular weight cutoff ranging from 2,900 to 13,000 could be formed [22] from chitosan. In 1984, Rha et al. first documented a procedure for preparing chitosan capsules for cell encapsulation [23]. The chitosan-alginate capsules had a liquid alginate core. Since then, several other studies have been reported on the use of chitosan copolymers for immobilization of hybridoma cells and plant cells [24,25]. However, the apparent poor biocompatibility of chitosan with hybridoma and insect cells was indicated by Smith et al. [26] and McKnight et al. [27].

This chapter focuses on various applications of chitosan, as well as current research on its physicochemical properties. Application areas that are covered include water treatment, pharmaceutics, biotechnology, food processing, and membranes.

PHYSICOCHEMICAL PROPERTIES OF CHITOSAN

Chitosan is a collective name given to a group of polymers deacetylated from chitin. The difference between chitin and chitosan lies in the degree of deacetylation. Generally, the reaction of deacetylating chitin in an alkaline solution cannot reach completion even under harsh treatment. The degree of deacetylation usually ranges from 70% to 95%, depending on the method used. These methods have been thoroughly reviewed by Muzzarelli [28]. The technique of Horowitz, for example, treating chitin with solid potassium hydroxide for 30 minutes at 180°C, results in the highest removal (95%) of acetyl groups. Recently, Kobayashi et al. [29] published a procedure for preparing chitosan from mycelia of absidia strains. A chitosan product with 79–91% deacetylation and 1,200,000 molecular weight was obtained. Most publications use the term chitosan when the degree of deacetylation is more than 70%.

Up to now, only a few studies on the molecular conformation of chitosan have been reported. There is still no one simple model describing chitosan, in spite of the fact that several models have been published for chitin, α-chitin [30] and β-chitin [31]. The α-chitin, for example, is tightly arranged in an antiparallel fashion, whereas β-chitin is in a parallel form. An analysis of the diffraction spectra of chitin and chitosan revealed a structural resemblance between the two polymers [28]. It has been suggested that the conformation of chitosan was similar to that of α-chitin [32].

Commercial chitosan is mainly produced by deacetylating chitin obtained from seashell materials. The quality and properties of chitosan products, such as purity, viscosity, deacetylation, molecular weight, and polymorphous structure, may vary widely because many factors in the manufacturing process can influence the characteristics of the final product. Kurita et al. [33], for example, found that the adsorption ability of chitosan for metal ions depended on the hydrolysis process. A homogeneous hydrolysis process could give a chitosan product with a higher adsorption rate than the one prepared by a heterogeneous process with the same degree of deacetylation. Furthermore, Bough et al. [34] investigated the influence of other manufacturing variables. They found that the highest viscosity and highest molecular weight chitosan product could be prepared by grinding the dry shrimp hulls to 1 mm prior to treatment, using alkali deproteination, purging nitrogen into the reaction vessel, and increasing the deacetylation time.

The degree of deacetylation is one of the more important chemical characteristics of chitosan. This determines the content of free amino groups in the polysaccharide. Methods for checking the removal of acetyl groups in chitosan include infrared spectroscopy [35,36], titration [37], gas chromatography [35], and dye adsorption [38] (Table 1). Among these, Muzzarelli suggested that first derivative ultraviolet spectrophotometry at 199 nm [39] was probably the best method for nondestructively and accurately determining the degree of acetylation in chitosan samples [40]. With this technique the *N*-acetylglucosamine absorbance readings were linearly dependent on concentration and were not influenced by the presence of acetic acid. Another method for analyzing the degree of deacetylation in chitosan was introduced by Maghami

Table 1. Physiochemical properties of chitosan.

Property	Comment	Reference
Degree of Deacetylation	Determined by:	
	UV spectrophotometry	39, 40
	dye adsorption	38
	IR spectrometry	35, 36
	metachromatic titration	37
	gas chromatography	35
Molecular Weight	Determined by:	
	chromatography	34
	light scattering	3
	viscometry	43
Viscosity	Affected by:	
	ionic strength	3, 44
	deacetylation time	34
	molecular weight	34, 44
	concentration	44
Solubility	Usually dissolved when pH < 6 but also affected by:	
	solvent mixing	28
	deacetylation	33
	solvation	47, 57
	chemical modification	45, 58
Coagulating Ability	Binding for:	
	metal ions	50–55
	anionic polymers	23–25, 55
	amino acids	59
	proteins	11, 12, 60, 61
	DNA	62, 63
	cells	13, 64, 65
	dyes	38, 66, 67
	solids	60, 68

and Roberts [38]. They found that under acid conditions there was a 1:1 stoichiometry for the interaction of amino groups in chitosan with sulfonic acid groups on dye ions. The dyeing of C. I. Acid Orange 7 was found to be the most rapid method for determining the deacetylation value in chitosan.

The molecular weight of native chitin is usually larger than one million while commercial chitosan products fall between 100,000 and 1,200,000. During the manufacturing process, harsh conditions can lead to degradation of the chitosan product. For example, with the Horowitz method, after a 30-minute treatment at 180°C, a chitosan sample with a chain length of only twenty units was obtained [28]. In general, factors such as dissolved oxygen, high temperature, and shear stress, can cause the further degradation of chitosan products. Dissolved oxygen can slowly degrade chitosan. At temperatures over 280°C thermal degradation of chitosan takes place and the polymer chains rapidly break down. On the other hand, shear degradation caused by hydrodynamic forces favors breakage of long-chain chitosan molecules to a critical length. Below this point, shearing does not further affect the molecular weight distribution [3]. This process can provide a relatively narrow molecular weight distribution. Several procedures for preparing low molecular weight chitosan have been reported. Takeda et al. [41], for example, obtained a colorless sample with an average molecular weight of only 60,000 by treating chitosan solution with 0.05% ClO_2. In contrast, Kushino and Orihara [42] treated a chitosan solution with enzymes such as papain, cellulase, and acid protease and obtained a product having an average molecular weight of 36,000.

The molecular weight of chitosan can be determined by methods such as chromatography [34], light scattering [3], and viscometry [43]. Among these, viscometry is the most simple and rapid method for determination of molecular weight. Although Bough et al. [34] indicated that the molecular weight of chitosan was not always directly related to its viscosity because of the presence of colloidal particles, the viscometry method is still widely employed for determining the relative molecular weight of chitosan. Maghami and Roberts [43] tested a series of chitosan samples having the same relative molecular mass distribution but different extents of *N*-acetylation by using the Mark-Houwink equation. Results showed that the equation was applicable to chitosan over the *N*-acetylation range from 0% to 40%.

The viscosity of chitosan in solution is influenced by many factors, such as the degree of polymer deacetylation, molecular weight, concentration, ionic strength, pH, and temperature (Table 1). In general, as the temperature rises, the viscosity of the polymer solution decreases.

However, a pH change in the polymer solution may give different results depending on the type of acids employed. With acetic acid, the viscosity of chitosan tends to increase with decreasing pH, whereas with HCl the viscosity decreases when the pH is lowered. Kienzle-Sterzer et al. [44] studied the persistence length of chitosan molecules in dilute solution. They indicated that the intrinsic viscosity of chitosan was a function of the degree of ionization as well as ion strength. An increase in either chitosan ionization or ion strength would decrease the polymer solution's intrinsic viscosity. Based on their study, they proposed that chitosan in dilute solution behaves as a non-draining worm-like molecule, its molecular configuration being dictated by the electrostatic interactions between polyion and counterions.

Chitosan is insoluble in water, alkali, and organic solvents but is soluble in most solutions of organic acids when the pH of the solution is less than 6. Acetic and formic acids are two of the most widely used acids for dissolving chitosan. Some dilute inorganic acids, such as nitric acid, hydrochloric acid, perchloric acid, and H_3PO_4, can also be used to prepare a chitosan solution but only after prolonged stirring and warming. Sometimes, however, a white gel-like precipitate is formed in nitric acid solution after polymer dissolution. In addition to dissolving chitosan in acid solution, mixtures such as dimethylformamide with dinitrogen tetroxide at a ratio of 3:1 also appear to be good solvents for chitosan [28].

Water-soluble chitosans, dissolved in the absence of acids, are frequently required when acids are undesirable substances in products, such as cosmetics, medicines, and foods. It has been shown that chitosan with 50% deacetylation from homogeneous processing is water-soluble [33]. The reason for this effect was found to be the presence of a small proportion of scarcely deacetylated macromolecules contained in the chitosan sample prepared by heterogeneous processes [45,46]. These molecules, which came from the more crystalline regions of the polysaccharide, tended to remain aggregated in solution and were more resistant to chemical reactions, whereas, under homogeneous conditions, the uniform distribution of N-acetyl groups decreased the polymer crystallinity and increased the solubility. Other methods for improving the water solubility of chitosan have been investigated. Kushino and Asano [47] reported a procedure for preparing a water-soluble chitosan salt. Chitosan was dissolved in an aqueous solution, concentrated (to 10%) by evaporation, and finally spray-dried at 175°C to provide water-soluble salt. Chemical modification of chitosan, on the other hand, provides an alternative to improve the biopolymer's water solubility. Based on the assumption that the carboxymethylation of chitosan can impart water solubility to the insoluble polysaccharide, Muzzarelli [45] prepared three water-soluble derivatives of chitosan, *O*-CM-chitosan, *N*-CM-chitosan and *N,O*-CM-chitosan. The *O*-CM-chitosan showed the best solubility over the pH range 3–11.

Finally, chitosan is a good coagulating agent and flocculant due to the high density of amino groups, which can interact with negatively charged substances, such as proteins, solids, dyes, and polymers. However, chitosan behaves quite differently with respect to transition metal ions. The nitrogen in the amino group of the chitosan molecule acts as an electron donor and is presumably responsible for selective chelating with metal ions. The complexation of the chitosan nitrogen with metal ions was confirmed by Tzsezos [48] and Ogawa and Oka [49]. Ogawa and Oka proposed that a metal ion (i.e., cupric ion) could coordinate with four amino groups in the D-glucosamine dimer residue of the chitosan chain. The free amino group in chitosan was considered to be much more effective for binding metal ions than the acetyl groups in chitin [50]. This leads us to consider that the higher free amino group content of chitosan should give higher metal ion adsorption rates. However, Kurita et al. [51] indicated that the adsorption ability of chitosan is dependent on many other factors, such as crystallinity, deacetylation, and affinity for water. In their studies, they found that samples with 55% deacetylation, prepared by a homogeneous hydrolysis process, showed the highest adsorption ability. These samples were amorphous and had the best solubility in water. The effect of structure on adsorption ability was also revealed. The ability of metal adsorption, for example, could be enhanced by cross-linking [52,53], controlled *N*-acetylation [54], and complexation with other polymers like glucan [55,56]. The chitosan-glucan complex demonstrated much stronger chelating ability than chitosan alone. Chitosan-glucan does recover several metal ions from water, including Cr, Co, Ni, Cu, Cd, and Pb.

APPLICATIONS OF CHITOSAN

The industrial production and use of chitosan has been steadily increasing since the 1970s. In Japan, for example, the production of chitosan increased 37% each year from 1978 to 1983, the total annual amount reaching 311 tons by 1983 [40] and 1,270 tons by 1986 [69]. At that time, the major applications of chitosan were centered on sludge dewatering, food processing, and metal ion chelation. The present trend, in industrial applications, however, is toward producing high value products, such as cosmetics, drug carriers, feed additives, semipermeable membranes, and pharmaceutics (Table 2). The difference in value between the products and the low-cost polymer is one of the main

Table 2. Applications of chitosan.

Applications	Examples	References
Water Treatment	Removal of Metal Ions	28, 40, 48–56, 72–74
	Flocculant/Coagulant:	
	Proteins	60–61
	Dyes	66–67
	Amino Acids	59
	Filtration	22, 75–76
Pulp and Paper	Surface Treatment	77–78
	Photographic Paper	79
	Carbonless Copy Paper	140
Medical	Bandages, Sponges	80–81
	Artificial Blood Vessels	82–85
	Blood Cholesterol Control	40, 86
	Tumor Inhibition	13, 87
	Membranes,	83–85
	Dental/Plaque Inhibition	58, 88–104
	Skin Burns/Artificial Skin	1, 91
	Eye Humor Fluid	89
	Contact Lens	1, 154
	Controlled Release of Drugs	25, 92–95, 155
	Bone Disease Treatment	90
Cosmetics	Make-up Powder	96
	Nail Polish	97
	Moisturizers	98
	Fixtures	99–102
	Bath Lotion	58, 103
	Face, Hand and Body Creams	58
	Toothpaste	58, 104
	Foam Enhancing	156
Biotechnology	Enzyme Immobilization	105–114, 141
	Protein Separation	11–12
	Chromatography	115, 157–158
	Cell Recovery	65, 159
	Cell Immobilization	23–24, 116–117, 147, 160
	Glucose Electrode	114
Agriculture	Seed Coating	20, 63, 118
	Leaf Coating	119–120
	Hydroponic/Fertilizer	74, 121–123
	Controlled Agrochemical Release	124
Food	Removal of Dyes, Solids, Acids	67,125–127
	Preservatives	128–131
	Color Stabilization	132
	Animal Feed Additive	74, 134
Membranes	Reverse Osmosis	3
	Permeability Control	28, 95, 146, 148–153
	Solvent Separation	161

driving forces pushing studies on new applications of chitosan. Biotechnology is currently attempting large-scale production of high-value bioproducts like monoclonal antibodies, which were projected at about 1.2 billion dollars (U.S.) on the world market in 1991 [70]. Immobilization techniques have been proven to be an effective way to increase cell density, product concentration, and hence, productivity in a culturing system. Chitosan membranes and gels have great potential for use in immobilized cell culture systems.

Water Treatment and Papermaking

One of the earliest applications of chitosan was for chelating harmful metal ions, such as copper, lead, mercury, and uranium, from wastewater. Chitosan is considered to be safe for human use [71]. Studies on the chelating ability of chitosan began about twenty years ago [72]. In 1973, Muzzarelli documented the effectiveness of chitin, chitosan, and other chelating polymers for their ability to chelate transition metal ions [28]. He indicated that chitosan was a powerful chelating agent and exhibited the best collection ability of all the tested polymers because of its high amino group content. Furthermore, Masri et al. [73] compared the chelating ability of chitosan with several other materials, such as bark, activated sewage sludge, and poly(p-aminostyrene) and confirmed that chitosan had a higher chelation ability than these substances. In another interesting study dealing with the recovery of uranium from water, Hirano et al. [74] showed that 40–74% of the available uranium could be recovered from river and lake water, and 3% from seawater. The extremely low recovery rate from seawater might have been caused by the presence of large quantities of other metal ions, which would have competed with uranium for complexing with chitosan. Recent studies showed that the chelating ability of chitosan could be further improved by homogeneous hydrolysis [51], cross-linking [52,53], controlled N-acetylation [54], and complexation with other polymers like glucan [55,56]. It was reported, for example, that the chelating capacity of cross-linked glycine chitosan for copper uptake was 22 times higher than that of chitosan itself, and 4 times higher than that of the chitosan-glucan complex [40].

Chitosan is an excellent coagulating agent and flocculant due to the high density of amino groups on the polymer chain that can interact with negatively charged substances, such as proteins, solids, and dyes. Wu et al. [60] investigated the effectiveness of different chitosans for removing proteins from cheese whey. They found that the effectiveness in coagulating solids and proteins was inversely proportional to the molecular weight of the polymer. The optimal concentrations for the chitosans tested ranged from 70 to 150 mg/mL, with a resultant reduction of more than 90% in the turbidity. In another area, adsorbants (prepared by mixing chitosan with cellulose fibers) were employed for sorption of dyes from wastewater [66]. The chitosan-cellulose fibers were able to adsorb 264 mg/g reactive dyes and 421 mg/g acid dyes compared with an adsorption capacity of less than 80 mg/g dye for coconut charcoal. Furthermore, No and Meyers [59] successfully used crawfish chitosan for the recovery of organic compounds and the coagulation of suspended solids from seafood-processing streams. In one study, they used two amino-Cu crawfish chitosan columns to recover amino acids from wastewater. The two columns were operated in turn to absorb the amino acids and elute Cu. Recoveries were reported to be 15% for glycine, 21% for proline, and 47–93% for the other amino acids.

Many types of chitosan membranes have been developed for water clarification and filtration. Nakajima and Shinoda [75] formed two types of membranes by reacting glycol chitosan with two polyanion mucopolysaccharides–chondroitin sulfate C and heparin. These membranes showed the lowest salt rejection at the isoelectric points of the mucopolysaccharides. Retention of 3–6% for urea, 54–57% for glucose, and 65–73% for sucrose was reported. Hirano et al. [22] prepared a series of chitosan membranes with a thickness from 12 to 60 μm by air drying. Depending on the solution, the membranes had permeation rates ranging from 2.7 to 340 kg per square meter per day. Most of them were stable in both dilute acid and alkali. Among the membranes formed, only the N-acetylchitosan membranes showed good semipermeable properties. The membranes were reported to be permeable to various compounds with molecular weights up to 2,900 and impermeable to cytochrome C of molecular weight 13,000. In another study, organic anions such as benzenesulfonate and benzoate ions appeared to be actively transported through chitosan-PVA membranes from the acid side to the alkaline side. The transport of amino acids occurred in the reverse order [76].

Several early applications of chitosan in the pulp and paper industries have been reviewed by Muzzarelli [77]. These include the surface treatment of paper with a 1% chitosan solution to increase its bursting strength and folding endurance, without any decrease in the paper's brightness. With the development of new technology, such as color photocopying, different types and higher quality paper and fibers are required. In the fiber industry, for example, the treatment with a 0.5% chitosan solution improved the fiber's color fastness [78]. In the paper making area, Aizawa and Noda [79] placed a chitosan layer on photographic paper to increase antistatic properties since electrostatic discharge can cause a serious decrease in picture quality. The surface resistance to electrostatic charging was reported to have increased more than ten thousand times after treatment with the chitosan solution.

Pharmaceutics and Biotechnology

Although chitosan by itself is hemostatic (stops bleeding), some derivatives, such as sulfated chitosan, are blood anticoagulants. By utilizing the hemostatic effect, chitosan bandages and sponges were prepared for surgical treatment and wound protection [80,81]. In contrast, Tokura et al. [82] immobilized partially sulfated chitosan oligomers on the surface of molded chitosan materials, such as artificial blood vessels and fibers, to produce antithrombogenic medical products. There are other medical applications that have also been investigated. Anticoagulant membranes for blood ultrafiltration can be prepared by immobilizing bioactive molecules like PGE1, hirudin, heparin, or antithrombin on chitosan [83–85]. Uragami and Mori [83] formed membranes by complexing quaternized chitosan with sodium heparin and found that the ultrafiltration rate of the membranes could be decreased by increasing the degree of heparinization. The in vivo tests showed that no thrombus formed on the membranes.

Chitosan has also been found to be effective against cholesterol. Various hypolipemic formulations containing chitosan, including particles, powders, solutions, and injections, were prepared for oral administration [86]. In oral tests on mice, these medicines effectively decreased the blood cholesterol level up to 66%. The hypocholesterolemic activity of chitosan was probably due to inhibition of micelle formation [40]. The latter contains cholesterol, as well as fatty acids and monoglycerides.

Sirica and Woodman [13] showed that chitosan could selectively aggregate leukemia tumor cells in vitro, producing a dense aggregate and inhibiting cell growth. In another study, after feeding gall mite *E. cladophthirus* with punctured *S. dulcamara* cells, Bronner et al. [62] observed that chitosan, which has a strong affinity for DNA, was transmitted from the mite to the nuclei of the punctured cells within one hour. In similar work, Hadwiger et al. [63] not only confirmed the interaction of chitosan with DNA, but they also built a model to explain the resistance response of host cells to the pathogen.

The special affinity of chitosan for biomolecules has been utilized to reduce side effects of drugs. For example, Ouchi et al. [87] covalently linked an antineoplastic prodrug, 5-fluorouracil, to chitosan with hexamethylene spacers via carbamoyl bonds. In mice tests, the chitosan bound prodrug exhibited an enhanced inhibition effect against tumor cells without displaying any apparent toxicity. The new drug was reported to be effective for treating leukemia by decreasing the tumor weight. It also showed a fivefold growth inhibition ratio for Meth-A fibrosarcoma over the prodrug itself.

The bacteriostatic action of chitosan was demonstrated by Allan et al. [1] against organisms selected from common skin bacterium. They found that *S. epidermis* was completely inhibited by a 0.1% solution of chitosan, but *S. aureus* and *P. aeruginosa* needed a 1% concentration. Shibasaki et al. [88] reported that the presence of low molecular weight chitosan in drinking water could inhibit the formation of plaque and cavities because of the inhibition of microbial growth. An eye humor fluid was also formulated by using *N*-carboxyacylchitosan that was used during eye surgery and treatment [89]. In this liquid, chitosan not only served to increase the solution viscosity, but also helped to prevent inflammation after surgery.

Chitosan, as a polymer, is a good additive in the preparation of biomaterials. In 1984, Allan et al. indicated that it may be possible to employ chitosan in creating soft contact lenses either by casting or molding procedures [1]. Contact lenses are required to be optically clear, safe, wettable, and gas permeable. Among these properties, gas permeability is particularly important because oxygen must be transported from the tear fluid to the eye surface, whereas carbon dioxide needs to diffuse back from the eye surface to the fluid. Several techniques for preparing contact lenses have been reported resulting in chitin *n*-butyrate lenses, chitosan lenses, and blue chitosan lenses [1].

Chitosan was also employed for making artificial skin [91]. The artificial chitosan dermis was tested by inserting it into a cut on the back of rats. A normal inflammatory reaction was observed after 2 days, followed by cell colonization after 7 days. Finally, low molecular weight chitosan (MW = 1,500) was found to be useful for treating bone disease. The low molecular weight polysaccharide effectively increased alkaline phosphatase activity, thereby accelerating bone formation [90].

Controlled-release technology is proving to be an effective means for delivering lower concentrations of drugs in the body and reducing side effects. Nagai et al. [92] and Tan et al. [93] used chitosan and its derivatives as additives with other materials such as lactose and starch, in the preparation of compressed tablets. The release of drugs from these tablets was found to be related in part to the loading of the chitosan additives and followed a zero-order profile. Bodmeier et al. [94] entrapped microparticles of drugs in chitosan beads formed by ionotropic gelation of chitosan in tripolyphosphate solution. On eluting with 0.1 N HCl, the chitosan beads disintegrated and released the microparticles. The disintegration time appeared to be a function of the polysaccharide viscosity, gelation time, and drying method. In another report, Kanke et al. [95] formed three types of chitosan films containing a model drug, prednisolone, for retarded release studies. The three types of films—monolayer, double-layer, and an N-acetylated chitosan double layer—were prepared by air drying. They found that the N-acetylated film gave a slower release rate than the other films.

Applications of chitosan in cosmetics were reviewed by Muzzarelli [77] several years ago. He indicated that the use of chitosan could help to remove leftover starch contained in shampoos. Its use would also have the effect of conferring shine and strength to hair due to interactions between the polysaccharide and hair proteins. Recently, more reports have been published on the use of chitosan and its derivatives in cosmetic products. Yabe et al. [96], for example, prepared a series of colored cosmetic powders for makeup by spray-drying a mixture of pigments and chitosan granules. The powders showed high fastness to sunlight and sweat. Other reports on the use of chitosan for cosmetics include nail polishes [97], moisturizers [98], hair and skin fixatives [99], and hair conditioners [100–102]. Cleansers can also contain chitosan. One of these, a bath lotion containing chitosan lactate, chitosan succinate, and chitosan alkyl phosphates, was prepared by Bandai et al. [103]. He claimed that this lotion resulted in increased skin softness.

Amino and hydroxy groups on the polymer provide places for physical or chemical linkages and modifications. Chitosan, therefore, has been widely used as a carrier for enzyme immobilization. Enzymes immobilized on chitosan gels include α-chymotrypsin [105,106], α-galactosidase [107], invertase [108], β-galactosidase [109], α-amylase [110], lactase [111], and cyclodextrin glucanotransferase [112]. The immobilization usually involved a cross-linking step using glutaraldehyde. In addition, Braun et al. [113] immobilized penicillin G acylase on chitosan powder and beads by a covalent linkage. They found that a covalently linked enzyme was active longer than an absorbed one. Another interesting application was by Katayama [114], who prepared a glucose electrode by forming an immobilized enzyme-chitosan membrane on a nylon net that was then covered with a hydrogen peroxide permeable membrane.

Several reports have been published on the use of chitosan as a polyligand, flocculating agent, and polymer support for separation and recovery of proteins and cells. Senstad and Mattiasson [11,12], for example, separated two proteins—trypsin and wheat germ agglutinin—from crude solutions by using chitosan affinity precipitation. The target proteins were first selectively bound to the chitosan molecules and then precipitated by raising the pH above 7. The precipitates were collected and redissociated, and the samples were finally purified with a gel filtration column. The recovery rates and the purification factors (specific activity of the enzyme after purification over specific activity before purification) were reported to be 66% and 5.5 for trypsin, and 70% and 10 for wheat germ agglutinin, respectively. A promising application of chitosan for affinity chromatography was reported by Holme et al. [115]. They immobilized 1-thio-β-glucopyranosyl on chitosan gels, which had a special affinity for β-D-glucosidase. The column was tested by loading A. faecalis β-D-glucosidase in a buffer containing 5 mM sodium phosphate at pH 6.8. The results showed that most of the enzyme was bound to the column. The bound enzyme was finally eluted by the addition of a salt solution containing 0.5 M NaCl and 5 mM sodium phosphate at pH 6.8.

Immobilization technology has the potential for providing high cell densities and high product concentrations in a cell culture system. Champluvier et al. [116] developed a procedure to immobilize yeast cells, *Kluyveromyces lactis* CBS 683, in a glass wool matrix treated with chitosan. Under the influence of various solvents, the adsorbed cells secreted intracellular β-galactosidase that was employed in the hydrolysis of lactose. Rha et al. [23] was apparently the first to attempt an application of chitosan in the encapsulation of living cells. In their procedure, capsules were formed by complexing chitosan with sodium alginate. In another study, Knorr et al. [24] encapsulated plant cells in chitosan-alginate and chitosan-carrageenan capsules for the production of secondary metabolites. They found that chitosan could effectively increase the concentration of oxalate in a cell culture medium. This phenomena was considered to be caused by the permeabilizing effect of chitosan. Yoshioka et al. [117] entrapped viable hybridoma cells in chitosan-CM-cellulose capsules and cultured them in a supplemented DMEM/F12 medium. A tenfold increase in cell density and a threefold higher product concentration were obtained in comparison to a suspended cell culture. For large-scale applications, however, the complexity of the encapsulation process and scale-up problems appear to be key problems.

Agriculture and Food Processing

Chitosan has many potential applications in agriculture because the polymer is essentially naturally occurring and biodegradable; therefore, it should not cause pollution problems. One application that is widely employed at present is seed coating. Hadwiger and his colleagues [63] found that chitosan treatment in coating seeds had many beneficial effects, such as inhibition of fungal pathogens in the vicinity of the seeds and enhancement of plant-resistant responses against diseases. By using this method, crop yields were increased from 10% to 30% for wheat, peas, and lentils. Because of the increase in crop yield, this method has been accepted for wheat coating in eleven states in the United States [118]. Other effects of chitosan treatment on plants have also been observed. Yano and Tsugita [20], for example, treated soybean seeds with 1% colloidal chitosan in 0.25% lactic acid. This resulted in a germination time of 3.1 days versus 4.1 days for the untreated controls. Pospieszny and Atabekov [119] sprayed a dilute chitosan solution on *Phaseolus vulgaris* bean leaves. This led to a significant reduction in the number of local lesions produced by the alfalfa mosaic virus. At present, chitosan is exempt from the requirement of a tolerance for residues under the Federal Food, Drug, and Cosmetic Act when the polysaccharide is used as a seed treatment for soybeans to accelerate emergence, to enhance early root and leaf growth, and to increase pod setting and yield [120].

In addition to directly treating plants with chitosan, the polysaccharide has also been employed for improving soil properties and preparing hydroponic fertilizer. For example, Hirano et al. [74] tested radish seedlings in an N-methylenechitosan gel bed and found that the water evaporation rate was greatly reduced by the gel bed. In another test, they mixed the N-methylenechitosan gels with sand and clay. This test showed that young seedlings in the mixture were still alive on the fourth day after watering had stopped, but seeds in sand withered within two days. In similar work, Takahashi [121,122] prepared a liquid amendment containing 25 mg/L chitosan for soil improvement. The amendment claimed to improve the granular structure of soils and to increase crop (i.e., strawberry) yield by 29%. In addition to soil improvement, chitosan can also be added to fertilizers. Recently, a preparation consisting of a liquid fertilizer, containing chitosan and trace elements such as Fe, Mn, Zn, Cu, and Mo, was employed for hydroponic culturing [123].

Concerns about the environment have aroused great interest in the excessive use of agrochemicals, such as fertilizers, herbicides, and pesticides. To reduce environmental damage caused by agrochemicals, the use of chitosan in controlled-release systems was investigated. Teixeira et al. [124] reported that by coating fertilizer pellets with films of N-acetyl or N-propionyl chitosan, a controlled-release effect could be achieved. In their studies, urea and a herbicide (atrazine) were trapped in either N-acetyl or N-propionyl chitosan gels. The studies showed that compared with uncoated control beads, the release period could be extended by a factor of 50 for atrazine beads and a factor of 180 for urea beads. The release time of atrazine was increased to six months in N-acetyl-chitosan beads and one year in N-propionyl-chitosan beads. These compare favorably to four days for uncoated control beads.

Owing to the high chelating and coagulating ability of chitosan, the polymer has been widely utilized in the food industry. In the case of beverages, for example, chitosan was used to remove dyes from orange juice [67] and to remove solids, β-carotene, and acid substances from apple and carrot juice [125–127]. In addition to the above, chitosan was also used to extend the preservation time of foods because of its anti-biotic properties. Haga and Enokida [128], for example, added 0.025% chitosan by weight to rice and steamed the mixture. They found that chitosan with molecular weight from 70,000 to 90,000 effectively controlled the growth of microorganisms for up to 3 days without adversely affecting the organoleptic (objective smelling test) scores. The chitosan treatment was also employed on other foods such as sardines [129], milk [130], and Chinese cabbages [131]. Other chitosan applications that have been reported include color stabilization of fish protein paste [132] and increased firmness of cucumber pickles [133].

Chitosan Membranes and Cell Encapsulation

Chitosan membranes may be prepared in various ways: evaporation of chitosan solvents, cross-linking with bifunctional reagents, chelating with anionic counterions, or complexing with polymers and proteins. The direct evaporation of a chitosan solution spread on a glass plate is the most simple technique for the preparation of chitosan films and generally produces a water-soluble film. This type of film was found to be amorphous and to have comparable water vapor permeability to cellophane [28]. In 1977, Muzzarelli [3] reported a procedure for the preparation of reverse osmosis membranes. These membranes were prepared by dissolving chitosan in a dilute acidic solution mixed with organic solvents, such as methanol, ethanol, and acetone, in a 6:4 ratio of water to solvent. The solution was spread on a glass plate followed by air drying. By this method a 3-μm-thick membrane was obtained which had a very high water permeability while rejecting 81% of the calcium chloride. Recently, Kanke et al. [95] compared two different types of chitosan films—monolayer and double-layer films. They found that the permeability of the double-layer films could be controlled by changing operation variables.

Cross-linked chitosan membranes can be prepared by the addition of bifunctional reagents, such as aldehydes, carboxylic anhydrides, and glutaraldehyde, to the chitosan solution. Hirano [21], for example, published a procedure for the preparation of N-arylidene-chitosan and N-acyl-chitosan membranes. A chitosan-hydroacetate solution was first poured into a glass petri dish to give a thin liquid layer. After the addition of either an aldehyde or carboxylic anhydride, films formed within a few hours. These membranes were reported to be neither solubilized nor swollen by soaking for one week in water, 2 N NaOH, dimethyl sulfoxide, or formamide. The mechanism of cross-linking gelation has been studied [135,136]. In Roberts and Taylor's study [136] of the gelation behavior of the chitosan-glutaraldehyde system, the effect of variables, including the activity coefficients of electrolytes and concentrations of chitosan, glutaraldehyde, and electrolytes, were examined. They suggested that the cross-linking mechanism involves formation of a Schiff's base structure.

Ionotropic gelation for the formation of chitosan membrane is a very mild process. Chitosan membranes have been formed with a variety of counterions or polymers, such as pyrophosphate, octylsulfate, and alginate. Among the counterion polymers that gelate with chitosan, alginate is the most widely used. Usually, alginate gels are prepared by reacting the polymer with divalent ions such as calcium. The encapsulation of living cells using alginate-polycations was developed by Lim and Sun [137] for transplantation purposes (bioartificial pancreas). In this process, a solution of sodium alginate and suspended cells was extruded into a calcium chloride bath. In the reaction that followed, sodium ions in the alginate solution were exchanged for calcium ions. The resultant gel beads were then reacted with a polycationic polymer [poly(L-lysine)] to form a membrane and hardened with polyethyleneimine (to reduce digestibility in vivo and capsule clumping). The hard inner core could be reliquefied with sodium citrate to increase the mass transfer rate [138]. In the membrane formation step, a salt bond formed between the positive amine on the polycationic polymer and the negative carboxylic acid on the alginate. Several modifications of this procedure have been attempted to increase the capsule's useful in vivo lifetime and membrane strength. Rha et al. [23] simplified the procedure of capsule formation and replaced poly(L-lysine) with chitosan. The formation of salt bonds between positively and negatively charged functional groups was used to explain the formation of the chitosan-alginate membrane. One advantage of Rha and Rodriguez-Sanchez's procedure [139] was direct membrane formation, which facilitated the free orientation of the polymer chains and increased the membrane strength.

Cell encapsulation technology is important in the development of new cell transplantation techniques for hormone delivery in medicine. One of the most important advantages of microencapsulation is the ability to provide a sheltered surrounding for the cells. The semipermeable capsule membrane allows small molecules to diffuse through but prevents the passage of large molecules and cells. Encapsulation may provide potentially new forms of treatment for a number of metabolic diseases of various organs. The transplantation of insulin-producing islets of Langerhans immobilized in alginate-polylysine microcapsules, for example, may be an effective new way of treating diabetes in animals [142].

Since the earliest industrial applications of encapsulation (the preparation of microcapsules containing ink for carbonless copy paper [140]), the technique has the potential for providing higher cell densities and product concentrations. A successful industrial application with multiple gram productions of monoclonal antibodies was reported by Posillico [144] and Rupp [145] using hybridoma cells encapsulated in alginate-polylysine membranes. Both the concentration and purity of the intracapsular product were approximately 100 times greater than that achieved in suspension cultures. In another study, a higher density of hybridoma cells was obtained in a modified multiple membrane capsule [143]. King et al. [146] further estimated that the density of insect cells immobilized in alginate-polylysine capsules with a multiple membrane was about 100 times greater than that achieved by traditional suspension culture. Chitosan-copolymer capsules have been successfully used to culture *Bacillus* [23], plant cells [24], and hyridoma cells [117,147]. These recent developments are summarized in Table 3.

The capsule membrane permeability and molecular weight cutoff can be controlled by modulating the viscosity average molecular weight and concentration of the membrane-forming polymers, pH, and ionic strength, as well as the reaction time [148,149,151–153]. The authors found that the durability of chitosan-alginate capsule membranes de-

Table 3. Whole-viable cell microencapsulation processes.

Membrane	Cell Type	Reference
Polyamides (Nylon)	Human red blood cells	141
Alginate-PLL-Polyethylene imine	Rat islets of Langerhans	137
Alginate-PLL-Alginate	Rat islets of Langerhans	142
		148
		149
		144
		145
Alginate-PLL-Alginate	Insect cells	146
Eudragit RL	Human red blood cells	162
Alginate-Eudragit RL	Human red blood cells	163
Polyacrylate	Mouse hybridoma cells	164
Cross-linked Albumin	Hybridoma cells	165
Chitosan-CM-Cellulose	Hybridoma cells	117
Chitosan-Alginate	Hybridoma cells	147, 148
Chitosan-Alginate	Microbial cells	23
Chitosan-Alginate	Plant cells	24
Chitosan-Carrageenan	Plant cells	24

pended on the chitosan molecular weight—the lower the chitosan molecular weight, the stronger and thicker the membranes. They postulated that this effect could be due to the limitation of the molecular size with respect to mass diffusion of chitosan through the pores in the alginate gel matrix.

CONCLUDING REMARKS

In the past thirty years, substantial progress has been made on fundamental and applied research in chitosan technology. One of the driving forces behind this rapid development has been the continuing decrease in the supply of natural resources. At the same time there has been an increased realization that there are abundant alternative bioresources. Two factors—economics and versatility—have also stimulated interest in chitosan's utilization in various fields.

It is clear that chitosan is a good material for the preparation of gels and films. It should not be difficult to prepare chitosan membranes with a specific permeability and molecular weight cutoff. These membranes have the potential for controlling the release of bioactive materials, such as drugs, herbicides, insecticides, fertilizers, and perfumes, and for innovating the design of membrane bioreactors.

Although beneficial characteristics of chitosan such as antifungal properties, the ability to coagulate with red cells, and the ability to complex with DNA have been shown, it is worthwhile to note that the mechanism of interaction between living cells and the functional groups on the cationic polymer still needs to be elucidated. In closing, we can safely state that chitosan is no longer just a waste by-product from the seafood-processing industry. This material is now being utilized by industry to solve problems and to improve existing products, as well as to create new ones.

REFERENCES

1. Allan, G. G., L. C. Altman, R. E. Bensinger, D. K. Ghosh, Y. Hirabayashi, A. N. Neogi and S. Neogi. 1984. In *Chitin, Chitosan and Related Enzymes*, J. P. Zikakis, ed., Academic Press, Inc., pp. 119–133.

2. Alper, J. 1984. *Science 84*, (March):36–43.

3. Muzzarelli, R. A. A. 1977. *Chitin*. Toronto: Pergamon of Canada Ltd.

4. Rigby, G. W. 1934. "Substantially Undergraded Deacetylated Chitin and Process for Producing the Same," U.S. Patent 2,040,879.

5. Rigby, G. W. 1934. "Process for the Preparation of Films and Filaments and Products Thereof," U.S. Patent 2,040,880.

6. Clark, G. L. and A. F. Smith. 1936. "X-ray Diffraction Studies of Chitin, Chitosan and Derivatives," *J. Phys. Chem.*, 40:863–879.

7. Bartnicki-Garcia, S. and E. Reyers. 1968. *Biochim. Biophys. Acta*, 165(1):32–42.

8. Hadwiger, L. A. and J. M. Backman. 1980. *Plant Physiol.*, 66(2):205–211.

9. Knorr, D. 1984. *Food Technology*, (January):85–97.

10. Brine, C. J. 1984. Chapters 17–23 in *Chitin, Chitosan and Related Enzymes*, J. P. Zikakis, ed., Academic Press, Inc.

11. Senstand, C. and B. Mattiasson. 1989. *Biotech. Bioeng.*, 34:387–393.

12. Senstand, C. and B. Mattiasson. 1989. *Biotech. Bioeng.*, 33:216–220.

13. Sirca, A. E. and R. J. Woodman. 1971. *J. Nat. Cancer Inst.*, 47(2):377–388.

14. Allan, C. R. and L. A. Hadwiger. 1979. *Exp. Mycol.*, 3(3):285–287.

15. Balassa, L. L. and J. F. Prudden. 1978. MIT Sea Grant Rep. MITSG, 78-7, *Proc. Int. Conf. Chitin/Chitosan*, 1st, PB 285 640, pp. 296–305.

16. Malette, W. G. Jr., H. J. Quigley and E. D. Adiches. 1986. In *Chitin in Nature and Technology*, R. Muzzarelli, C. Jeuniaux and G. W. GooDay, eds., New York: Plenum Press, pp. 435–442.

17. Suzuki, K., Y. Ogawa, K. Hashimoto, S. Suzuki and M. Suzuki. 1984. *Microbiol. Immunol.*, 28:903–912.

18. Nishimura, K., S. Nishimura, N. Nishi, S. Tokura and I. Azuma. 1986. In *Chitin in Nature and Technology*, R. Muzzarelli, C. Jeuniaux and G. W. GooDay, eds., New York: Plenum Press, pp. 477–483.

19. Eida, T. and H. Hidaka. 1988. *Jpn. Fudo Saiensu*, 27(12):56–63.

20. Yano, S. and T. Tsugita. 1988. "Chitosan-Containing Seed Coatings for Yield Enhancement," Jpn. Kokai Tokkyo Koho JP 63,139,102 [88,139,102].

21. Hirano, S. 1978. *Agric. Biol. Chem.*, 42(10):1939–1940.

22. Hirano, S., K. Tobetto, M. Hasegawa and N. Matsuda. 1980. *J. Biomedical Materials Research*, 14:477–486.

23. Rha, C., D. Rodriguez-Sanchez and C. Kienzle-Sterzer. 1984. In *Biotechnology of Marine Polysaccharides*, R. R. Colwell, E. R. Pariser and A. J. Sinskey, eds., Washington: Hemisphere Publishing Corp., pp. 283–311.

24. Knorr, D., M. D. Beaumont and Y. Pandya. 1987. *Biotechnol. Food Ind., Proc. Int. Symp.*, pp. 389–400.

25. Shiotani, T. and Y. Shiiki. 1986. "Preparation of Capsules Using Phosphates and Chitin Derivatives," Jpn. Kokai Tokkyo Koho JP 61,153,135 [86,153,135].

26. Smith, N. A., M. F. A. Goosen, G. A. King, P. Faulkner and A. J. Daugulis. 1988. *Biotechnology Letters*.

27. McKnight, C. A., A. Ku, M. F. A. Goosen, D. Sun and C. Penney. 1987. "Synthesis of Chitosan-Alginate Microcapsule Membranes," *J. Bioactive and Compatible Polymers*, 3:334–355.

28. Muzzarelli, R. A. A. 1973. *Natural Chelating Polymers*. Toronto: Pergamon of Canada Ltd., pp. 83–95.

29. Kobayashi, T., Y. Takiguchi, K. Shimahara and T. Sannan. 1988. *Nippon Nogei Kagaku Kaishi,* 62(10):1463–1469.

30. Prasad, N. and C. Ramakrishnan. 1972. *Indian J. Pure Appl. Phys.,* 10:501–505.

31. Gardner, K. H. and J. Blackwell. 1975. *Biopolymers,* 14:1581–1595.

32. Averbach, B. L. 1975. Report MITSG 75-17, NOAA 75102204, National Information Service, U.S. Department of Commerce.

33. Kurita, K., T. Sannan and Y. Iwakura. 1977. *Makromol. Chem.,* 178:3197–3202.

34. Bough, W. A., W. L. Salter, A. C. M. Wu and B. E. Perkins. 1978. *Biotech. Bioeng.,* 20:1931–1943.

35. Muzzarelli, R. A. A., F. Tanfani, G. Scarpini and G. Laterza. 1980. *J. Biochem. Biophys. Methods,* 2:299–306.

36. Domszy, J. G. and G. A. F. Roberts. 1985. *Macromol. Chem.,* 186:1671–1677.

37. Gummow, B. D. and G. A. F. Roberts. 1985. *Macromol. Chem.,* 186:1239–1244.

38. Maghami, G. G. and G. A. Roberts. 1988. *Makromol. Chem.,* 189:2239–2243.

39. Muzzarelli, R. A. A. 1984. In *New Development in Industrial Polysaccharides,* V. Crescenzi and I. C. M. Dea, eds., New York: Gordon and Breach, pp. 417–450.

40. Muzzarelli, R. A. A. 1985. In *The Polysaccharides,* G. O. Aspinall, ed., London: Academic Press, Inc., 3:417–451.

41. Takeda, K., K. Shimizu and M. Goto. 1989. *Manufacture of Low Molecular Weight Chitosan,* Jpn. Kokai Tokkyo Koho JP 01,11,101 [89,11,101].

42. Kushino, S. and Y. Orihara. 1988. "Enzymatic Manufacture of Low Molecular-Weight Chitosan and Its Use in Pharmaceutical, Food, and Other Manufacture," Jpn. Kokai Tokkyo Koho JP 63,63,388 [88,63,388].

43. Maghami, G. G. and G. A. F. Roberts. 1988. *Makromol. Chem.,* 189:195–200.

44. Kienzle-Sterzer, C., D. Rodriguez-Sanchez and C. Rha. 1984. In *Chitin, Chitosan and Related Enzymes,* J. P. Zikakis, ed., Academic Press, Inc., pp. 383–396.

45. Muzzarelli, R. A. A. 1988. *Carbohydrate Polymers,* 8:1–21.

46. Rinaudo, M. and A. Domaro. 1989. In *Chitin and Chitosan,* G. Skjak-Braek, T. Anthonsen and P. Sanford, eds., New York: Elsevier Applied Science, pp. 71–86.

47. Kushino S. and H. Asano. 1988. "Water-Soluble Chitosan," Jpn. Kokai Tokkyo Koho JP 63,225,602 [88,225,602].

48. Tsezos, M. 1983. *Biotech. Bioeng.,* 25:2025–2040.

49. Ogawa, K., K. Oka, T. Miyanishi and S. Hirano. 1984. In *Chitin, Chitosan and Related Enzymes,* J. P. Zikakis, ed., Academic Press, Inc., pp. 327–346.

50. Maruca, R., B. J. Suder and J. P. Wightmen. 1982. *J. Appl. Polym. Sci.*, 27:4827–4837.

51. Kurita, K., T. Sannan and Y. Iwakura. 1979. *J. Appl. Polym. Sci.*, 23:511–515.

52. Koyama, Y. and A. Taniguchi. 1986. *J. Applied Polymer Sci.*, 31:1951–1954.

53. Kurita, K., Y. Koyama and A. Taniguchi. 1986. *J. Applied Polymer Sci.*, 31:1169–1176.

54. Kurita, K., S. Chikaoka and Y. Koyama. 1988. *Chem. Letters*, pp. 9–12.

55. Muzzarelli, R. A. A., F. Tanfani and G. Scarpini. 1980. *Biotech. Bioeng.*, 22:885–896.

56. Muzzarelli, R. A. A., F. Tanfani, G. Scarpini and E. Tucci. 1980. *J. Applied Biochem.*, 2:54–59.

57. Lang, G. and T. Clausen. 1989. In *Chitin and Chitosan*, G. Skjak-Braek, T. Anthonsen and P. Sandford, eds., New York: Elsevier Applied Science, pp. 139–147.

58. Muzzarelli, R. A. A. 1989. In *Chitin and Chitosan*, G. Skjak-Braek, T. Anthonsen and P. Sandford, eds., New York: Elsevier Applied Science, pp. 87–99.

59. No, H. K. and S. P. Meyers. 1989. *J. Food Sci.*, 54(1):60–62.

60. Wu, A. C. M., W. A. Bough, M. R. Holmes and B. E. Perkins. 1978. *Biotech. Bioeng.*, 20:1957–1968.

61. No, H. K. and S. P. Meyers. 1989. *J. Agric. Food Chem.*, 37(3):580–583.

62. Bronner, R., E. Westphal and F. Dreger. 1989. *Physiol. Mol. Plant Pathol.*, 34(2):117–130.

63. Hadwiger, L. A., B. Fristensky and R. C. Riggleman. 1984. In *Chitin, Chitosan, and Related Enzymes*, J. P. Zikakis, ed., Academic Press, Inc., pp. 291–302.

64. Horisberger, M. and M. F. Clerc. 1989. *Histochemistry*, 90(3):165–175.

65. Gualtieri, P., L. Barsanti and V. Passarelli. 1988. *Ann. Inst. Pasteur/Microbiol.*, 139(6):717–726.

66. Shinagawa, K., G. Takemura and A. Kobayashi. 1979. Jpn. Kokai Tokkyo Koho, Pat. 79,152,685.

67. Seo, T., T. Kanbara and T. Iijima. 1988. *J. Appl. Polym. Sci.*, 36(6):1443–1451.

68. Watanabe, I. and H. Bandai. 1988. "Cleaning Agents for Bathtubs," Jpn. Kokai Tokkyo Koho JP 63,12,699 [88,12,699].

69. Hirano, S. 1989. In *Chitin and Chitosan*, G. Skjak-Braek, T. Anthonsen and P. Sandford, eds., New York: Elsevier Applied Science, pp. 37–43.

70. McCormick, D. 1989. *Bio/Technology*, 5:27.

71. Hirano, S., H. Seino, Y. Akiyama and I. Nonaka. 1989. *Polym. Mater. Sci. Eng.*, 59:897–901.

72. Muzzarelli, R. A. A. and O. Tubertini. 1969. *Talanta*, 16:1571–1577.

73. Masri, M. S., F. W. Reuter and M. J. Friedman. 1974. *Applied Polymer Sci.*, 18:675–681.

74. Hirano, S., H. Senda, Y. Yamamoto and A. Watanabe. 1984. In *Chitin, Chitosan, and Related Enzymes*, J. P. Zikakis, ed., Academic Press, Inc., pp. 77–95.

75. Nakajima, A. and K. Shinoda. 1977. *J. Appl. Pol.*, 21:1249.

76. Uragami, T., F. Yoshida and M. Sugihara. 1988. *Sep. Sci. Technol.*, 23(10–11):1067–1082.

77. Muzzarelli, R. A. A. 1983. *Carbohydrate Polymers*, 3:53–75.

78. Tokunaga, M. 1988. "Improvement of Colorfastness of Cellulosic Fibres and Their Blends Dyed with Direct and/or Reactive Dyes," Jpn. Kokai Tokkyo Koho JP 63,175,186 [88,175,186].

79. Aizawa, Y. and T. Noda. 1988. "Antistatic Photographic Paper," Jpn. Kokai Tokkyo Koho JP 63,189,859 [88,189,859].

80. Kibune, K., Y. Yamaguchi, K. Motosugi. 1988. "Manufacture of Bandages from Chitin Fibres," Jpn. Kokai Tokkyo Koho JP 63,209,661 [88,209,661].

81. Motosugi, K., Y. Yamaguchi and K. Kibune. 1988. "Chitosan Sponges as Surgical Dressings," Jpn. Kokai Tokkyo Koho JP 63,90,507 [88,90,507].

82. Tokura, S., M. Itoyama and S. Hiroshi. 1988. "Partially Sulfated Chitosan Oligomers Immobilized on Chitosan for Antithrombogenic Medical Goods," Jpn. Kokai Tokkyo Koho JP 63,89,167 [88,89,167].

83. Uragami, T., H. Mori and Y. Noishiki. 1988. *Jinko Zoki*, 17(2):511–514.

84. Chandy, T. and C. P. Sharma. 1988. *Polym. Sci. Technol.*, 38:297–311.

85. Chandy, T. and C. P. Sharma. 1989. *J. Colloid Interface Sci.*, 130(2): 331–340.

86. Suzuki, S., M. Suzuki and H. Katayama. 1988. "Chitin and Chitosan Oligomers as Hypolipemics and Formulations Containing Them," Jpn. Kokai Tokkyo Koho JP 63,41,422 [88,41,422].

87. Ouchi, T., T. Banba, T. Matsumoto, S. Suzuki and M. Suzuki. 1989. *J. Bioact. Comp. Polymers.*

88. Shibasaki, K., T. Matsukubo, N. Sugihara, E. Tashiro, Y. Tanabe and Y. Takaesu. 1988. *Koku Eisei Gakkai Zasshi*, 38(4):572, 573.

89. Miyata, T., T. Yoneda and M. Izume. 1988. "Manufacture of Eye Humour Fluid Substitutes with *N*-carboxyacylchitosan," Jpn. Kokai Tokkyo Koho JP 63,220,865 [88,220,865].

90. Namita, K., N. Miyajima, M. Higo and J. Nakayama. 1988. "Pharmaceuticals Containing Chitin, Chitosan, and Their Derivatives for Treatment of Bone Diseases," Jpn. Kokai Tokkyo Koho JP 63,156,726 [88,156,726].

91. Collumbel, C., O. Damour, C. Gagnieu, F. Poinsignon, C. Echinard and J. Marichy. 1988. "Biomaterials for Artificial Skin and Implants Containing Acetylated Chitosan, Collagens, and Glycosaminoglycans," Eur. Pat. Appl. EP 296,078.

92. Nagai, T., Y. Saway and N. Nambu. 1984. In *Chitin, Chitosan, and Related Enzymes*, J. P. Zikakis, ed., Academic Press, Inc., pp. 21–40.

93. Tian-Rui, T., K. Inouye, Y. Machida, T. Sannan and T. Nagai. 1988. 48(4):318–321.

94. Bodmeier, R., H. Chen and O. Paeratakul. 1989. *Pharm. Res.*, 6(5):413–417.

95. Kanke, M., H. Katayama, S. Tsuzuki and H. Kuramoto. 1989. *Chem. Pharm. Bull.*, 37(2):523–525.

96. Yabe, H., Y. Kawamura and I. Kurahashi. 1988. "Coloured Cosmetic Bases Containing Fine Powders of Chitin and Chitosan," Jpn. Kokai Tokkyo Koho JP 63,161,001 [88,161,001].

97. Lang, G., G. Maresch and H. R. Lenz. 1989. "Nail Polishes Containing *O*-Benzyl-*N*-Hydroxyalkylchitosan as Film-Forming Component," Ger. Offen. DE 3,723,811.

98. Matsumura, S., T. Karigome and K. Nomoto. 1989. "Manufacture of *N*-Carboxymethoxycarbonylchitosan for Moisturizers," Jpn. Kokai Tokkyo Koho JP 01,14,203 [89,14,203].

99. Lang, G. and H. Wendel. 1988. "Hair and Skin Fixative Based on Chitosan and on Ampholytic Copolymer," Ger. Offen. DE 3,644,097.

100. Lang, G. and H. Wendel. 1988. "Quaternized Chitosan Derivatives, A Method for Their Preparation, and Cosmetics Containing Them," Eur. Pat. Appl. EP 290,741.

101. Mizushima, M. 1988. "Hair Preparations Containing Surfactants with Chitin or Chitosan Derivatives as Stabilizing Agents," Jpn. Kokai Tokkyo Koho JP 63,165,307 [88,165,307].

102. Iwao, S. 1989. "Hair Preparations Containing Polyoxyalkylene and Water-Soluble Chitin and Chitosan Derivatives for Protecting Hair," Jpn. Kokai Tokkyo Koho JP 01,09,911 [89,09,911].

103. Bandai, H., I. Watanabe and K. Murayama. 1988. "Bath Preparations Containing Chitosan or Its Derivatives," Jpn. Kokai Tokkyo Koho JP 63,10,715 [88,10,715].

104. Komiyama, N. and H. Itoi. 1988. "Denture Cleaners Containing Water-Soluble Chitins and Chitosans," Jpn. Kokai Tokkyo Koho JP 63,267,361 [88,267,361].

105. Kise, H., A. Hayakawa and H. Noritomi. 1987. *Biotech. Lett.*, 9(8): 543–548.

106. Muzzarelli, R. A. A., A. Barontini and R. Rocchetti. 1976. *Biotech. Bioeng.*, 18:1445–1454.

107. Ohtakara, A., M. Mitsutomi and Y. Uchida. 1986. In *Chitin in Nature and Technology*, R. Muzzarelli, C. Jeuniaux and G. W. GooDay, eds., New York: Plenum Press, pp. 409–411.

108. Illanes, A., R. Chamy and M. E. Zuniga. 1986. In *Chitin in Nature and Technology*, R. Muzzarelli, C. Jeuniaux and G. W. GooDay, eds., New York: Plenum Press, pp. 411–415.

109. Sicsic, S., J. Leonil, J. Braun and F. LeGoffic. 1986. In *Chitin in Nature and Technology*, R. Muzzarelli, C. Jeuniaux and G. W. GooDay, eds., New York: Plenum Press, pp. 420–422.

110. Synowiecki, J. 1986. In *Chitin in Nature and Technology*, R. Muzzarelli, C. Jeuniaux and G. W. GooDay, eds., New York: Plenum Press, pp. 417–420.

111. Van Griethuysen, E., E. Flaschel and A. Renken. 1986. In *Chitin in Nature and Technology*, R. Muzzarelli, C. Jeuniaux and G. W. GooDay, eds., New York: Plenum Press, pp. 422–428.

112. Nakakuki, T., M. Yoshida and M. Okada. 1988. "Immobilization of Cyclodextrin Glucanotransferase on Chitosan Beads," Jpn. Kokai Tokkyo Koho JP 63,196,290 [88,196,290].

113. Braun, J., C. P. Le and G. F. Le. 1989. *Biotechnol. Bioeng.*, 33(2):242–246.

114. Katayama, H. 1988. "Enzyme Membrane Electrodes Using Chitosan Membrane-Immobilized Enzymes," Jpn. Kokai Tokkyo Koho JP 63,78,062 [88,78,062].

115. Holme, K. R., L. D. Hall, C. R. Armstrong and S. G. Withers. 1988. *Carbohydrate Research*, 173:285–291.

116. Champluvier, B., F. Marchal and P. G. Rouxhet. 1989. *Enzyme Microb. Technol.*, 11(7):422–430.

117. Yoshioka, T., R. Hirano, T. Shioya and M. Kako. 1990. *Biotech. Bioeng.*, 35:66–72.

118. Sandford, P. A. and G. P. Hutchings. 1987. In *Industrial Polysaccharides, Progress in Biotechnology*, M. Yalpani, ed., 3:363–376.

119. Pospieszny, H. and I. G. Atabekov. 1989. *Plant Sci.*, 62(1):29–31.

120. 1989. United States Environmental Protection Agency, "Poly-D-Glucosamine (Chitosan); Exemption from the Requirement of a Tolerance," Fed. Regist. 23 Mar. 1989, 54(55):11948–11949 (Eng.).

121. Takahashi, N. 1988. "Soil Amendments Containing Poly Vinyl Alcohol, Chitosan and Tannins." 1988. Jpn. Kokai Tokkyo Koho JP 63,205,388 [88,205,388].

122. Takahashi, N. 1988. "Addition of Chitosan to Soils for Improvement of the Physical Properties," Jpn. Kokai Tokkyo Koho JP 63,146,982 [88,146,982].

123. Nowosielski, O., H. Struszczyk, T. Skwarski, S. Kotlinski, K. Krajewski and W. Dziennik. 1988. "Liquid Hydroponic Fertilizer Containing Chitosan and Trace Elements," Pol. PL 141,381.

124. Teixeira, M., W. P. Paterson, E. J. Dunn, Q. Li, B. K. Hunter and M. F. A. Goosen. 1990. *Ind. Eng. Chem. Res.*, 29:1205–1209.

125. Horman, I. 1980. Demande Brevet Fr. Pat. 2,435,917.

126. Imeri, A. G. and D. Knorr. 1988. *J. Food Sci.*, 53(6):1707–1709.

127. Soto-Peralta, N. V., H. Mueller and D. Knorr. 1989. *J. Food Sci.*, 54(2):495, 496.

128. Haga, M. and S. Enokida. 1988. "Chitosan-Containing Preservatives and

Sterilants for Foods," Jpn. Kokai Tokkyo Koho JP 63,169,975 [88,169,975].

129. Asao, Y., K. Sakuma and T. Machama. 1989. "Preservatives Containing Chitosan and Ascorbic Acid for Seafoods," Jpn. Kokai Tokkyo Koho JP 01,71,439 [89,71,439].

130. Kin, M., K. Kikuchi, I. Yamamoto and N. Yagi. 1988. "Milk Protein-Chitosan Complexes for Kamaboko and Other Products," Jpn. Kokai Tokkyo Koho JP 63,169,939 [88,169,939].

131. Izume, M. and Y. Uchida. 1988. "Food Preservation and Its Enhancement with Chitosan and/or Chitosanase-Digestion Products," Jpn. Kokai Tokkyo Koho JP 63,251,072 [88,251,072].

132. Wakayama, Y. and Y. Sekino. 1987. "Stabilization of Colouring Agents and Coloured Products with Chitosan," Jpn. Kokai Tokkyo Koho JP 62,297,365 [877,297,365].

133. Kuwahara, Y., N. Otsuka and M. Manabe. 1988. *Nippon Shokuhin Kogyo Gakkaishi*, 35(11):776–780.

134. Kono, M., T. Matsui and C. Shimizu. 1987. *Nippon Suisan Gakkaishi*, 53(1):125–129.

135. Moore, G. K. and G. A. F. Roberts. 1980. *Int. J. Biol. Macromol.*, 2:115.

136. Roberts, G. A. F. and K. E. Taylor. 1989. *Makromol. Chem.*, 190(5): 951–960.

137. Lim, F. and A. M. Sun. 1980. *Science*, 210:908–910.

138. Lim, F. 1982. "Encapsulation of Biological Materials," U.S. Patent 4,352,883.

139. Rha, C. and D. Rodriguez-Sanchez. 1988. "Encapsulated Active Material System," U.S. Patent 4,749,620.

140. Green, B. K. 1955. "Pressure-Sensitive Record Materials," U.S. Patent 2,712,507.

141. Chang, T. M. S. 1964. *Science*, 146:524, 525.

142. Sun, A. M. and G. M. O'Shea. 1985. *J. Control. Rel.*, 2:137–141.

143. King, G. A., A. J. Daugulis, P. Faulkner and M. F. A. Goosen. 1987. *Biotech. Progress*, 3(4):231–240.

144. Posillico, E. G. 1986. *Biotechnol.*, 4(2):114–117.

145. Rupp, R. G. 1985. In *Large-Scale Mammalian Cell Culture*, J. Feder and W. R. Tolbert, eds., Academic Press, Inc., pp. 19–38.

146. King, G. A., A. J. Daugulis, P. Faulkner, D. Bayly and M. F. A. Goosen. 1989. *Biotech. Bioeng.*, 34:1085–1091.

147. Kim, S. K. and C. Rha. 1989. In *Chitin and Chitosan*, G. Skjak-Braek, T. Anthonsen and P. Sandford, eds., New York: Elsevier Applied Science, pp. 617–626.

148. Goosen, M. F. A., G. A. King, C. A. McKnight and N. Marcotte. 1989. *J. Membrane Science*, 41:323–343.

149. Goosen, M. F. A. 1987. In *CRC Critical Reviews in Biocompatibility*, D. F. Williams, ed., Boca Raton, FL: CRC Press, p. 1.

150. Daly, M. M. and D. Knorr. 1988. *Biotech. Progress*, 4(2):76–81.

151. Daly, M. M., R. W. Keown and D. W. Knorr. 1989. "Chitosan Alginate Capsules," U.S. Patent 4,808,707.

152. Shioya, T. and C. Rha. 1989. In *Chitin and Chitosan*, G. Skjak-Braek, T. Anthonsen and P. Sandford, eds., New York: Elsevier Applied Science, pp. 627–634.

153. Kim, S. K. and C. Rha. 1989. In *Chitin and Chitosan*, G. Skjak-Braek, T. Anthonsen and P. Sandford, eds., New York: Elsevier Applied Science, pp. 627–642.

154. Markey, M. L., L. M. Bowman and M. V. W. Bergamini. 1989. In *Chitin and Chitosan*, G. Skjak-Braek, T. Anthonsen and P. Sandford, eds., New York: Elsevier Applied Science, pp. 713–717.

155. Brine, C. J. 1989. In *Chitin and Chitosan*, G. Skjak-Braek, T. Anthonsen and P. Sandford, eds., New York: Elsevier Applied Science, pp. 679–691.

156. Poole, S. 1989. In *Chitin and Chitosan*, G. Skjak-Braek, T. Anthonsen and P. Sandford, eds., New York: Elsevier Applied Science, pp. 523–531.

157. Roberts, G. A. F. and K. E. Taylor. 1989. In *Chitin and Chitosan*, G. Skjak-Braek, T. Anthonsen and P. Sandford, eds., New York: Elsevier Applied Science, pp. 577–583.

158. Seo, H. and Y. Kinemura. 1989. In *Chitin and Chitosan*, G. Skjak-Braek, T. Anthonsen and P. Sandford, eds., New York: Elsevier Applied Science, pp. 585–588.

159. Holland, C. R. 1989. In *Chitin and Chitosan*, G. Skjak-Braek, T. Anthonsen and P. Sandford, eds., New York: Elsevier Applied Science, pp. 559–566.

160. Nilsson, K. and K. Mosbach. 1980. *FEBS Lett.*, 118(1):145–150.

161. Uragami, T. 1989. In *Chitin and Chitosan*, G. Skjak-Braek, T. Anthonsen and P. Sandford, eds., New York: Elsevier Applied Science, pp. 783–792.

162. Sefton, M. V. and R. L. Broughton. 1982. *Biochimica et Biophys. Acta*, 717:473–477.

163. Lamberti, F. V., M. A. Wheatley, R. A. Evangelista and M. V. Sefton. 1983. *Am. Chem. Soc., Polym. Prepr.*, 24(1):75, 76.

164. Gharapetian, H., M. Maleki, N. A. Davies and A. M. Sun. 1986. *Polym. Mater. Sci. Eng.*, 54:114–118.

165. Tice, T. R. and W. E. Meyers. 1985. Eur. Pat. Application, No. 0,129,619.

CHAPTER 2

Applications of Chitin and Chitosan in the Ecological and Environmental Fields

SHIGEHIRO HIRANO
Department of Agricultural Biochemistry and Biotechnology
Tottori University
Tottori 680, Japan

INTRODUCTION

Chitin is a $(1\rightarrow4)$-linked 2-acetamido-2-deoxy-β-D-glucan, and chitosan is an N-deacetylated derivative of chitin. Chitin is the main component in the cuticles of crustaceans, insects, and mollusks and in the cell walls of some microorganisms (Muzzarelli, 1973). Chitin is hydrolyzed by both chitinase and lysozyme, and chitosan is hydrolyzed by chitosanase (Figure 1). Chitin is now produced commercially from crab and shrimp shells by treatment with dilute NaOH solution for deproteinization, followed by treatment with dilute HCl solution for demineralization. Chitosan is produced by treatment of chitin with concentrated NaOH at elevated temperature.

This chapter describes environmentally and ecologically friendly uses of chitin, chitosan, and their derivatives. Potential applications are also presented.

MOLECULAR CHARACTERISTICS OF CHITIN AND CHITOSAN

Chitin and chitosan are 1) the main components of crab and shrimp shells, which are abandoned by processing companies of marine products; 2) naturally occurring rare cationic aminopolysaccharides; 3) bio-

31

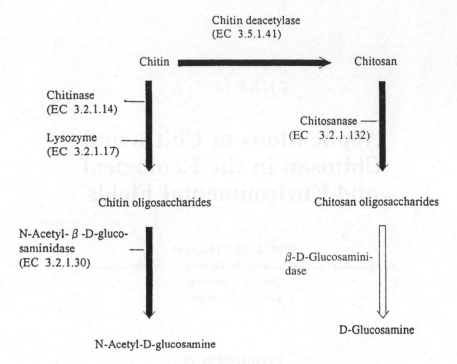

Figure 1. Enzymes for the hydrolysis of chitin and chitosan.

logically reproducible on the earth; 4) biodegradable on the earth; 5) biocompatible with organs, tissues, and cells of animals and plants; 6) almost nonantigenic in animal tissues and organs; 7) almost non-toxic in oral and implant administrations; 8) able to be processed into several casting products including flakes, fine powders, beads, membranes, sponges, cottons, fibers, and gels; 9) functional physically at high viscosity, moisturizing, metal chelating, polyelectrolyte forming, and affinity binding; 10) functional biologically in organs, tissues, and cells of animals and plants (the biosphere), and in the soil-, atmo- and hydrospheres; 11) definite in their chemical structures; and 12) modifiable chemically and enzymatically.

CURRENT COMMERCIAL USES OF CHITIN AND CHITOSAN

Several articles have dealt with wide uses of chitin, chitosan, and their derivatives in various fields (Knorr, 1984; Hirano, 1986a, 1986b, 1989a; Nihon Kogyokai, 1987; Jap. Soc. Chitin/Chitosan, 1990, 1995). Table 1 summarizes current practical applications.

Table 1. Current practical uses of chitin, chitosan, and their derivatives.

Uses	Compounds
Cationic sludge dewatering and flocculating agents for polluted wastewaters	Chitosan
Recovery of metal ions and proteins in aqueous waste solutions	Chitosan
Agricultural materials (e.g., plant seed coating, fertilizer)	Chitin and chitosan
Food additives	Chitin and chitosan
Feed additives for pets, fishes, and animals	Chitin and chitosan
Food processing (e.g., in sugar refining)	Chitosan
Hypocholesterolemic agents	Chitosan
Dressing materials for the burns and skin lesions of humans and animals and for plant tissue wounds	Chitin and chitosan
Biomedical materials (e.g., adsorbable suture)	Chitin, N-acylchitosans
Blood anticoagulant materials (e.g., heparinoids)	Sulfated chitin
Blood antithrombogenic materials	N-Hexanoyl and N-octanoyl derivatives
Hemostatic materials	Chitosan
Cosmetic ingredients for hair and skin cares	Chitosan, CM-chitin, HP-chitin
Textile and woven fabrics	Chitin, chitosan, chitin xanthate
Paints and dyeing and weaving	Chitosan
Natural thickeners	Chitosan
Papers, films, sponge materials	Chitin and chitosan
Chromatographic and immobilizing media	Chitin, chitosan, N-acylchitosan N-alkylidenechitosan
Analytical reagents (e.g., colloid titration and enzyme substrates)	Chitosan, HP-chitosan, HP-chitin, CM-chitin, CM-chitosan, N-acetylchitosan, D-glucosamine, N-acetyl-D-glucosamine, chitin and chitosan oligosaccharides

ENVIRONMENTALLY AND ECOLOGICALLY FRIENDLY APPLICATIONS

Enhancement of Biological Self-Defense Function

Immunological reactions in animal bodies are well known as a biological self-defense function against disease infections. Chitin, chitosan, and their derivatives enhance immunoadjuvant (Nishimura et al., 1984, 1985, 1986, 1987), antitumor (Suzuki et al., 1986), and antiviral and antibacterial functions (Iida et al., 1987; Tokoro et al., 1989) in animals. In plants, chitin, chitosan, and their derivatives enhance the induction of various biological self-defense compounds, including phytoalexins (pisatin, phaseolin, rishitin, orchinol, ipomeamarone, etc.) (Yamada et al., 1993), pathogenesis-related (PR) proteins (Schlumbaum et al., 1986; Inui et al., 1991), protein inhibitors (Lorito et al., 1994), and lignins (Notsu et al., 1994). Chitinase activity in plant seeds is enhanced during their natural germination and seeding stages because of their own biological self-defense function (Figure 2). The activity is

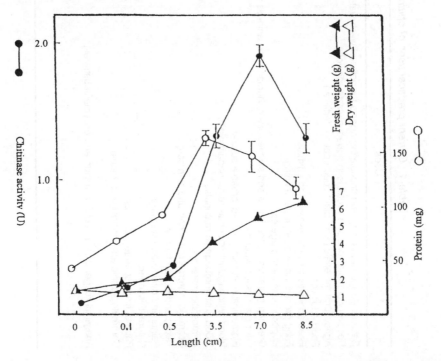

Figure 2. The enhancement of chitinase and PAL activities in the rice callus (3 g) treated with 200 μl (●) of CM-chitin dissolved at a concentration of 500 μg/ml in a 20 mM sodium phosphate buffer solution (pH 6.0), and only with the buffer solution as control (○).

Table 2. *Enhancement of extracellular lysozyme activity in culture media of mammalian cells in response to chitin, chitosan, and their derivatives.*[a]

Cultured Mammalian Cells	Extracellular Lysozyme Activity					
	A	B	C	D	E	F
Chicken embryo fibroblast cells	+ + +	+			+ +	
Monkey kidney vero cells			+ + +	+ + +	+ +	
Rat vascular smooth muscle cells			+ +		+ +	+ + +
Dog skin fibroblast cells			+ + +			

[a]A, chitin; B, chitosan; C, CM-chitin; D, HE-chitin; E, chitin oligosaccharides (d.p. 2–8); F, chitosan oligosaccharides (d.p. 2–10). Relative activity: + + +, high; + +, middle; + weak, and the blank parts are not determined.

enhanced up to 1.5 times by coating the seeds with chitosan (Notsu et al., 1994; Hirano et al., 1990), and the treatment prevents microbial disease infections and increases the plant production (Hadwiger et al., 1984; Hirano et al., 1988; Tsugita, 1995). In response to chitin, chitosan, and their derivatives, animal cells induce extracellular lysozyme (Table 2), and plant cells induce extracellular chitinase and phenylalanine ammonia-lyase (PAL) (Notsu et al., 1994) (Figure 3). These enzymes hydrolyze pathogen cell walls and prevent microbial infections in plants and animals.

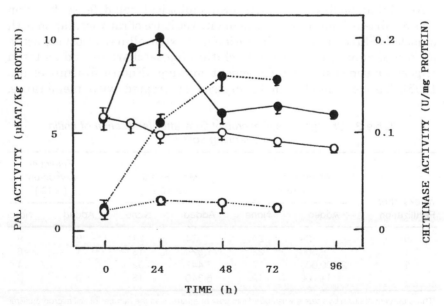

Figure 3. Changes in the chitinase activity, protein content, and length of the seedlings of Japanese black pine seeds during the germination stage.

Improvement of Microbial Flora in the Soilsphere and in Animal Intestines

Chitin and chitosan are fertilized into farming soils, and the soil microbial flora are improved. The total number of the useful microorganisms (e.g., *Actinomycetes*) increases, and that of harmful microorganisms (e.g., *Fusarium*) decreases (Table 3). Chitosan inhibits the growth of both *Fusarium* and *Helminthoporium* in an in vitro test, because of the ionic binding of these pathogens on the surface of cationic chitosan chains (Table 4) (Hirano et al., 1989b). The growth of *Escherichia coli* is inhibited almost completely in the presence of higher than 0.025% chitosan in cultured media (Uchida, 1988). Figure 4 shows several commercial products for ecologically friendly agricultural uses.

Chitosan feeding in rabbits improves their intestinal bacterial flora (Austin et al., 1981). The digestibility of chitosan is 35% on the fifth day after the chitosan feeding starts and increases to 80% on the fifteenth day. The increase in digestibility is due to an increase in the number of chitosanase-secreting bacteria in the intestines (Hirano et al., 1990a, 1990b).

Wound Healing Acceleration

All living bodies have their own biological self-defense function against disease infections. In the in vitro culture of rat vascular smooth muscle (VSM) cells, cell proliferation and extracellular lysozyme activity are enhanced in response to chitin oligosaccharides and chitosan oligosaccharides in a d.p.-dependent manner (Figure 5) (Inui et al., 1995). Tissue wounds can be dressed or treated with membranes,

Table 3. Changes in soil microbial flora after fertilization of chitin into farming soils.[a]

Weeks after Fertilization	Actinomycetes ($\times 10^4$)		Mold Fungi ($\times 10^4$)		Fusarium oxysporum ($\times 10^2$)	
	Added	None	Added	None	Added	None
0	24	24	24	24	9	9
1	8,500	22	1,528	13	1	6
2	2,000,000	17	7,447	32	0	4
3	11,700	12	5,340	15	0	3

[a]Three percent of chitin by volume was added into soils of a farm, and the number of soil microorganisms per cubic centimeter in the soils was counted.

Table 4. Growth inhibition of phytopathogenic fungi in response to chitosan and its oligosaccharides.[a]

	Growth in Percent of Control	
Fungi	Chitosan	Chitosan Oligosaccharides[b]
Alternaria alternata	69 ± 7	94 ± 3
Botrytis cinerea	81 ± 7	91 ± 4
Fusarium oxysporum sp. melonis	56 ± 9	69 ± 2
Fusarium oxysporum sp. lycopersici	51 ± 9	69 ± 2
Helminthosporum oryzae	45 ± 9	93 ± 4

[a]Circular portion (diameter 5 mm for the fungus mycelia) was inoculated in the center of the potato dextrose agar (PDA) medium containing a test compound (1.0 mg/ml) in a petri dish. For the inoculation of spores, the spore suspension (20 μl each) was put on the center of the PDA medium. The plates were incubated at 25°C for 3–7 days until the colony diameter was in the range from 5 to 6 cm. The longest and shortest diameters for each of the five colonies were measured.
[b]d.p. 2–9.

sponge sheets, cottons, fine powders, solutions and pastes, which are made of chitin, chitosan, or their derivatives. In wounds, chitinase activity is enhanced (Figure 6) (Hirano, 1993), tissue growth is stimulated, and the wound healing is accelerated by inhibition of microbial infection (Hirano et al., 1994a).

These data indicate that chitin, chitosan, and their derivatives are usable as a new wound-healing material. Several wound dressings as an artificial skin have been manufactured from chitin and chitosan and are commercialized for the healing of both human and animal wounds (Figure 7) (Kifune, 1995).

Chitosan is also usable as a dental cavity-preventing material. Low molecular weight (LMW) chitosan inhibits adhesion of *Streptococcus sorbrinus* on a hydroxyapatite granule coated with human saliva (Sano et al., 1991). Chitosan has a buffer capacity at a neutral pH range in human saliva and in human dental plaques (Shibasaki et al., 1994). Chitosan exhibits dental anticaries in rats infected with *Streptococcus sorbrinus* (Shibasaki et al., 1995). The mineralization of newly formed bone tissues and the periodontal healing are accelerated in response to chitosan and its derivatives via a chitosan-metal chelate (Muzzarelli et al., 1994). More clinical data are required for the practical uses.

Biocompatibility and Absorbability in Plant and Animal Tissues

Chitin, chitosan, and their derivatives are used as a biocompatible and absorbable material in both animal and plant tissues (Hirano, 1993). *N*-Hexanoyl and *N*-octanoyl derivatives of chitosan are compat-

Figure 4. Commercial products of chitin and chitosan for agricultural uses. Pellets of the blended composites of powdered crab shells and fish meals (left). Powdered crab shells (upper right) and aqueous solution of chitosan and its oligosaccharides (lower right).

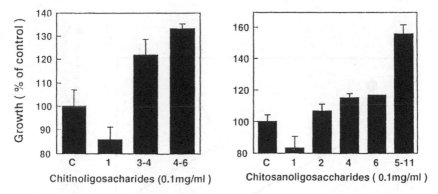

Figure 5. Stimulation of cell proliferation in culture medium of rat VSM cells treated with chitin oligosaccharides or chitosan oligosaccharides.

ible with animal bloods and tissues but are not digestible by lysozyme. However, N-acetyl (C2) to N-butyryl (C4) derivatives are digestible (Hirano et al., 1987). The data indicate that N-hexanoyl and N-octanoyl derivatives of chitosan are usable as an antithrombogenic material for contact lenses and artificial blood vessels.

Chitinase and its isoforms are distributed widely in tissues, organs, and body fluids of plants (Hirano et al., 1988), animals, microorganisms, and insects (Ohtakara, 1995). Plant and animal cells induce chitinase or lysozyme in response to chitin, chitosan, or their derivatives as an elicitor. Under tree bark tissues, the chitin membrane is digested within 3 months, and the N-propionylchitosan membrane within 5 months, but the chitosan membrane is not digested even after 7 months (Figure 8) (Hirano, 1993). The digestibility is controlled by the structure of N-substituents and by their substituent degrees (Hutadilok et al., 1995). Chitin, chitosan, and their derivatives are usable as an absorbable, implantable material in animal and plant tissues for controlled release of drugs and as a carrier material for the targeting of drugs to specific cells, tissues, or organs (Miyazaki et al., 1988, 1990; Song et al., 1993). A commercial product is a chitin suture absorbable in human and animal tissues. It is not necessary to take out the suture after a clinical operation because the suture is digested by lysozyme in the tissues (Figure 9).

Preservation of Biological Freshness

The freshness of vegetables and fruits is preserved by coating them with chitosan (Ghaouth et al., 1991, 1992a, 1992b, 1992c, 1994a,

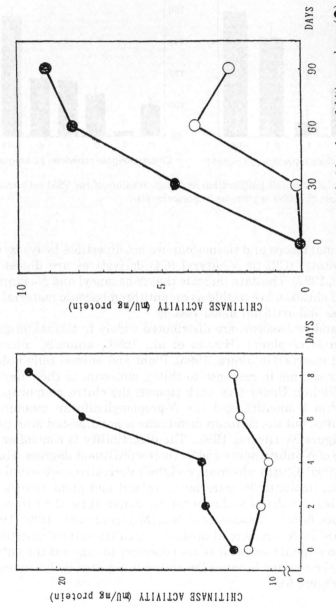

Figure 6. Enhancement of chitinase activity in bark tissue wounds that were covered with a sheet of chitin membrane (●) and without any cover as a control (○) in two trees, *Camellia japonica* L. var. *trifida Makino* (left) and *Prunus sargentii Rehd* (right).

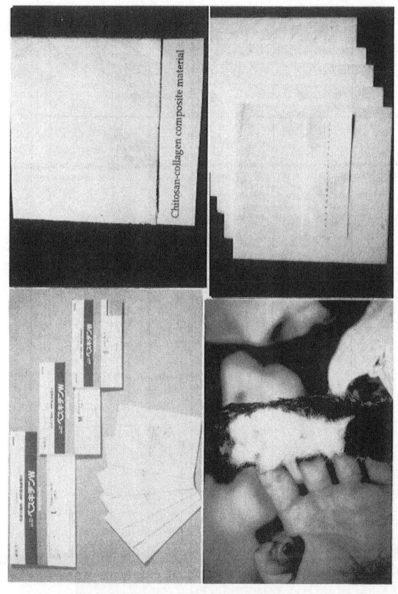

Figure 7. Commercial wound dressings of chitin and chitosan for human, animals, and trees. A nonwoven fabric dressing of chitin (upper left) and composite dressing of chitosan and atero-collagen (upper right) for human tissue wounds. Chitosan cottons (lower left) for animal tissue wounds and nonwoven composite sheets of chitin and cellulose for tree tissue wounds (lower right).

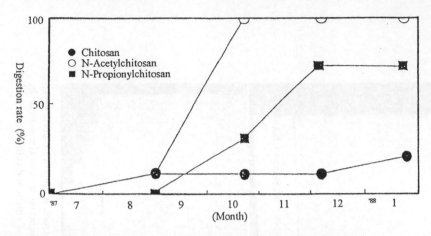

Figure 8. Fate of the membranes of chitin, *N*-propionylchitosan, and chitosan, implanted under tree bark, *Chamaecyparis pisifero* var. *Plimosa aurea*.

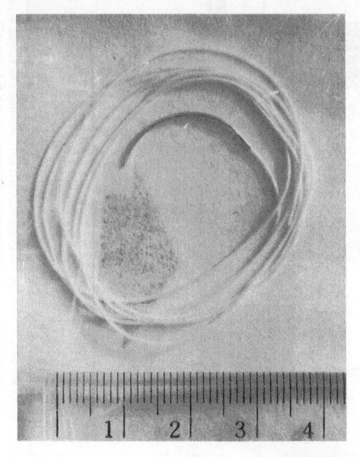

Figure 9. View of an absorbable chitin suture for humans and animals.

1994b). The chitosan coating prevents the release of CO_2 and ethylene from fruits and vegetables, and delays their ripening and microbial infections (Figure 10), resulting in the extension of the storage life of fruits and vegetables (Ghaouth et al., 1992d).

Functionality as an Additive for Feeds and Foods

Orally administered chitin and chitosan are digested by chitinase and chitosanase, which are secreted from intestinal bacteria. The digestibility of orally administered chitin and chitosan is almost 100% in hens and broilers. The digestibility of chitin is only 30% in rabbits, and the digestibility is almost unchanged on the fifth to twenty-fifth day after the feeding starts. However, the digestibility of orally administered chitosan is 35% on the fifth day, and the digestibility increases to 80% on the fifteenth day because of an increase in chitosanase-secreting bacteria in rabbit intestines (Hirano et al., 1990a, 1990b).

Hypocholesterolemic activity is observed with animals by the oral administration of chitosan. As shown in Table 5, the serum cholesterol level is decreased (Sugano et al., 1978; Kobayashi et al., 1979; Nagyvary et al., 1979; Sugano et al., 1980; Vahouny et al., 1983; Furda, 1983; Sugano et al., 1988; Fukada et al., 1991; Hirano and Akiyama, 1995), and the ratio of HDL-cholesterol/total cholesterol increases after the oral administration of chitosan. Orally administered chitosan is dissolved in the gastric acidic juice and is coagulated in the intestinal alkaline fluid to form micelles. The micelles entrap cholic acid as an ionic salt, and the circulation of cholic acid from the intestines to the liver is prevented. Cholesterol in the blood is utilized for supplying of intestinal cholic acid, resulting in a decrease of blood cholesterol levels. The micelles are digested by bacteria in the large intestines.

Serum lysozyme activity is enhanced after i.v. injection of chitosan oligosaccharides at a dose of 4.5 mg/kg body weight per day for 7–11

Table 5. Effects of 2% chitosan-supplemented diet feeding on the serum and liver cholesterol levels of rabbits fed a 0.9% cholesterol-enriched diet.[a]

Diet	Feeding Period (days)	Serum (ng/dl)				Liver (mg/g)		Liver Weight (g)
		Total CH	HDL-CH	TG	FFA	Total CH	TG	
A	0	79 ± 4	37	140 ± 8	0.06			
	39	650 ± 210	53	320 ± 62	0.12	14 ± 2	12 ± 2	97 ± 17
B	0	76 ± 12	32	120 ± 7	0.06			
	39	300 ± 130	58	210 ± 40	0.10	8 ± 2	8 ± 2	9 ± 21

[a]A, a 0.9% cholesterol-enriched diet; B, a 2% chitosan-supplemented and 0.9% cholesterol-enriched diet.

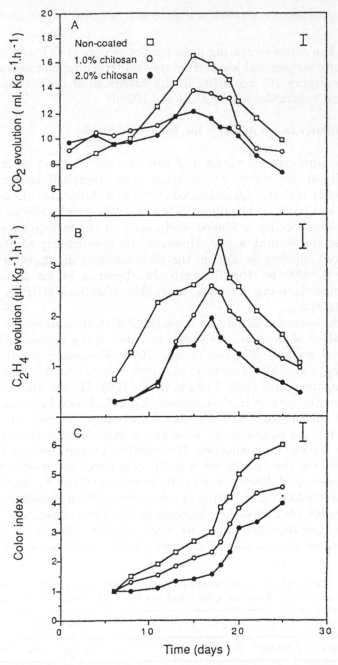

Figure 10. Effect of chitosan on respiration and ethylene production and quality of tomatoes during storage stage. Carbon dioxide evolution (A), ethylene production (B), and changes in ripening color index (1 = mature green, and 6 = red) (C) on uncoated (□) and 1% (○) and 2% (●) chitosan-coated tomatoes stored at 20°C.

days, resulting in prevention of microbial infection (Hirano et al., 1991b).

These data indicate that chitosan is usable not only as a biocompatible material but also as a biologically functional additive for feeds and foods. The hypocholesterolemic function of chitosan in the intestines differs from that of clinically used Chlestyramin resin, which is indigestible in the intestines. Figure 11 shows some commercial chitinous products of environmentally and ecologically functional feeds and foods.

Polyelectrolyte and Chelate Formations

Chitosan reacts with both polyanionic polymers and metal ions to produce precipitates as polyelectrolyte and chelate complexes (Shinoda et al., 1975; Kikuchi, 1975; Hirano et al., 1978). These reactions are usable for the clarification of polluted wastewater and for the recovery of proteins and metal ions from industrial wastewater. In 1975, a Japanese company first introduced chitosan-acetate salt as a natural cationic flocculating and dewatering agent for wastewater treatment. The system is still used for the recycling of used water (e.g., for swimming pools), the recovery of proteins and metals from industrial wastewater, the collection of bioactive compounds from animal and human urines, and the removal of endotoxins from biological extracts (Hashimoto, 1995). Chitosan is also usable as an adsorbent for the removal of harmful radioisotopes from water (Muzzarelli et al., 1972), and chitosan and its phosphate derivatives are usable as an adsorbent for the recovery of uranium ions from seawater (Sakaguchi et al., 1979) and from freshwater (Hirano et al., 1982).

Molecular Physical Functions

Chitosan, CM-chitin (*O*-carboxymethylchitin) and HE-chitin (*O*-hydroxyethylchitin) have unique physical properties including the moisturization of human skin, protection against mechanical damage to hair, antielectrostatic function, antimicrobial function, and prevention of skin aging. These properties of chitin, chitosan, and their derivatives are usable as a functional ingredient for cosmetics and as a functional material for textile fabrics (e.g., underwear). Figure 12 shows some ecologically friendly commercial cosmetics and fabrics.

Mineralization of CO_3^{2-} Ions in the Hydrosphere

To mimic mineralization of CO_3^{2-} in crab and shrimp shells, chemical mineralization can be attempted on the surface of chitosan mem-

Figure 11. Commercial products of foods and feeds containing chitosan as an additive: Japanese noodles (upper), a soybean sauce and a soybean paste (lower left), and canned cat feed (lower right).

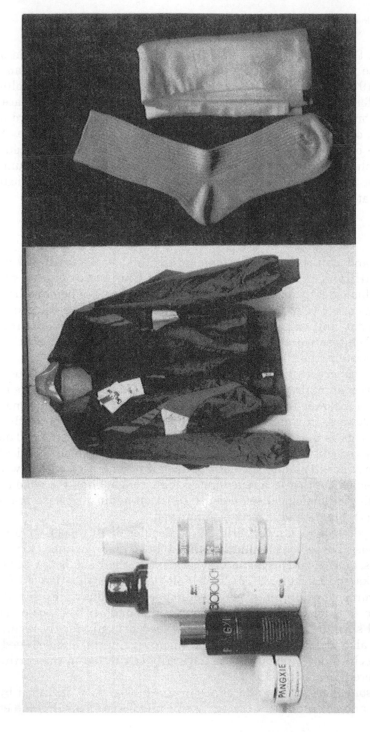

Figure 12. Ecologically friendly commercial products for cosmetics and fabrics. Cosmetics for skin and hair care (left), a windbreaker woven with a chitosan-coated synthetic layer (middle), and fabrics woven with a cellulose-chitin composite (99:1, w/w).

47

branes. A thin layer of chitosan solution in aqueous acetic acid on a glass plate is air dried to give a thin transparent chitosan membrane (Hirano, 1978). The membrane is dipped into an aqueous $CuCl_2$ solution to afford a bright green membrane of a chitosan-$(CuCl_2)_n$ chelate. The membrane is dipped into an aqueous K_2CO_3 solution to afford a dark blue color membrane of a chitosan-$[CuCO_3 \cdot Cu(OH)_2]_n$ chelate. A composite of alginate-$CaCl_2$ gel with chitin is treated in an aqueous K_2CO_3 solution to afford an alginate-$(CaCO_3)_n$-chitosan compound (Hirano et al., 1994c, 1995). The method is usable for the chemical mineralization of CO_3^{2-} and HCO_3^- ions in seawater (pH 8.2).

CONCLUSIONS

Chitin and chitosan are environmentally and ecologically active polymers (Hirano, 1991b; Hirano et al., 1994a). Chitin and chitosan are produced biologically on the earth at an estimated rate of one hundred billion tons every year. All of this is biologically degraded at the same time without any excess accumulation. This natural circulation of chitin and chitosan on the earth (Figure 13) results in the conservation of the ecosystem and the environment.

In the biosphere, chitin and chitosan are digestible in fluids, tissues, and organs of animals and plants (Hirano et al., 1994a). A butterfly sits on petals of a flower for a short period to take a drop of sweet honey from the flowers. During this short period, a portion of the butterfly's chitinous legs is digested by chitinase secreted from the petals. The depolymerized chitin and chitosan stimulate the induction of biological defense enzymes in plants. The damaged leg is regenerated with newly synthesized chitin to give a regenerated leg stronger than the old one. As a result, the biological self-defense function against disease infections of both plants and insects is enhanced.

In the hydrosphere, atmospheric CO_2 molecules dissolve as CO_3^{2-} and HCO^{3-} ions. These ions are mineralized by crabs, shrimps, krills, shellfishes, and calcareous algae (e.g., corals) into their shells as $CaCO_3$ (Borowitzka and Larkum, 1976). This biological mineralization of CO_2 in the hydrosphere prevents the greenhouse effect due to excess atmospheric CO_2 on the earth. Chitinous compounds in the bodies of crabs and shrimps turn into a functional feed not only for these animals but also for other fish, plankton, microorganisms, and seaweed. Thus, the chitinous compounds are naturally circulated in the earth's hydrosphere (Hirano, 1993).

In the soilsphere, chitin and chitosan are biologically degraded by chitinase and chitosanase secreted by soil microorganisms (Hirano et

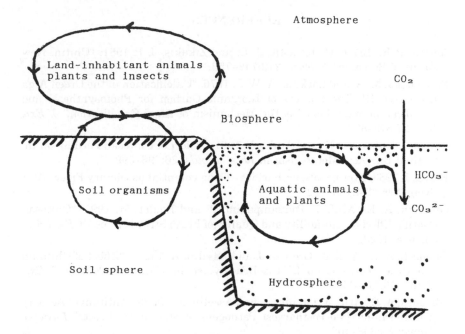

Figure 13. Natural circulation of chitinous compounds in the hydro-, bio-, soil-, and atmospheres on the earth. Chitinous compounds of insects and fungi are circulating in the soil- and atmospheres, and those of crustaceans and mollusks are in the hydrosphere. Atmospheric CO_2 is dissolved as CO_3^{2-} and HCO_3^- ions into the hydrosphere. These ions are mineralized as $CaCO_3$ in shells of crabs, calcareous algae, and shellfishes.

al., 1991a). The depolymerized products are in contact with plant roots on eluting with rainwater, and the biological self-defense function of plants is enhanced, resulting in an increase in plant productions.

Chitin and chitosan are neither medicines, foods, feeds, fertilizers, insecticides, nor fungicides. As described above, they are ecologically and environmentally bioactive natural polysaccharides (Hirano et al., 1994a). These polysaccharides should be used without disturbing their natural circulation. They should be used to help conserve the earth's environment and ecosystem.

ACKNOWLEDGEMENTS

This work was sponsored by the New Energy and Industrial Technology Development Organization (NEDO), and the Research Institute of Innovative Technology for the Earth (RITE), Kyoto, Japan.

REFERENCES

Austin, P. R., Brine, C. J. Castle, J. E. and Zikakis, J. P. 1981. "Chitin: New Facets of Research," *Science,* 212:749–753.

Borowitzka, M. A. and Larkum, A. W. D. 1976. "Calcification in the Green Alga *Halimeda.* III. The Sources of Inorganic Carbon for Photosynthesis and Calcification and a Model of the Mechanism of Algal Calcification," *J. Exp. Bot.,* 27:879–893.

Fukada, Y., Kimura, K. and Ayaki, Y. 1991. "Effect of Chitosan Feeding on Intestinal Bile Acid Metabolism in Rats," *Lipids,* 26:396–399.

Furda, I. 1983. "Aminopolysaccharides–Their Potential as Dietary Fiber," *ACS Symp. Ser.,* 214:105–122.

Ghaouth, A. E., Arul, J., Ponnampalam, R. and Boulet, M. 1991. "Chitosan Coating Effect on Storability and Quality of Fresh Strawberries," *J. Food Sci.,* 56:1618–1620.

Ghaouth, A. E., Arul, J., Grenier, J. and Asselin, A. 1992a. "Effect of Chitosan and Other Polyions on Chitin Deacetylase in *Rhizopus stolonifer,*" *Exp. Mycol.,* 16:173–177.

Ghaouth, A. E., Arul, J., Grenier, J. and Asselin, A. 1992b. "Antifungal Activity of Chitosan on Two Postharvest Pathogens of Strawberry Fruits," *Phytopathology,* 82:398–402.

Ghaouth, A. E., Arul, J., Asselin, A. and Benhamou, N. 1992c. "Antifungal Activity of Chitosan on Post-harvest Pathogens: Induction of Morphological and Cytological Alterations in *Rhizopus stolonifer,*" *Mycol. Res.,* 96:769–779.

Ghaouth, A. E., Ponnampalam, R., Castaigne, F. and Arul, J. 1992d. "Chitosan Coating to Extend the Storage Life of Tomatoes," *Hort. Sci.,* 27:1016–1018.

Ghaouth, A. E., Arul, J., Wilson, C. and Benhamou, N. 1994a. "Ultrastructural and Cytochemical Aspects of the Effect of Chitosan on Decay of Bell Pepper Fruit," *Physiol. Mol. Plant Pathol.,* 44:427–432.

Ghaouth, A. E., Arul, J., Grenier, J., Benhamou, N., Asselin, A. and Belanger, R. 1994b. "Effect of Chitosan on Cucumber Plants: Suppression of *Pythium aphanidermatum* and Induction of Defense Reactions," *Cyt. Histol.,* 84:313–320.

Hadwiger, L. A., Fristensky, R. and Riggleman, R. C. 1984. "Chitosan, a Natural Regulator in Plant-Fungal Pathogen Interactions, Increases Crop Yields," in *Chitin, Chitosan, and Related Enzymes,* J. P. Zikakis, ed., Orlando: Academic Press, p. 291.

Hashimoto, M. 1995. "Waste-water Treatments," (in Japanese), in *A Handbook of Chitin and Chitosan,* Jap. Soc. Chitin/Chitosan, ed., Tokyo: Gihodo Publishing Co., Inc., pp. 485–504.

Hirano, S. 1978. "A Facile Method for the Preparation of Novel Membranes from *N*-Acyl and *N*-Arylidene-chitosan Gels," *Agric. Biol. Chem.,* 42:1939–1940.

Hirano, S., Mizutani, C., Yamaguchi, Y. and Miura, O. 1978. "Formation of the

Polyelectrolyte Complexes of Some Acidic Glycosaminoglycans with Partially *N*-Acylated Chitosans," *Biopolymers,* 17:805–810.

Hirano, S., Noishiki, Y., Kinugawa, J., Higashijima, H. and Hayashi, T. 1987. "Chitin and Chitosan for Use as a Novel Biomedical Material," in *Advances in Biomedical Polymers*, Gebelein, C. G., ed. New York: Plenum Press, pp. 285–297.

Hirano, S., Kondo, Y. and Nakazawa, Y. 1982. "Uranylchitosan complexes," *Carbohydr. Res.,* 100:431–434.

Hirano, S. 1986a. "Chitin and Chitosan," in *A Handbook of Recent Biomedical Materials and Their Application* (in Japanese), Seno, M. and Otsubo, O., eds. Tokyo: R&D Planning, pp. 235–245.

Hirano, S. 1986b. "Chitin and Chitosan," *Ullmann's Encycl. Ind. Chem.,* A6:231–232.

Hirano, S., Hayashi, M., Murae, K., Tsuchida, H. and Nishida, T. 1988. "Chitosan and Derivatives as Activators of Plant Cells in Tissues and Seeds," in *Applied Bioactive Polymeric Materials,* Gebelein, C. G., Carraher, Jr., G. C. and Foster, V. R., eds. New York: Plenum Press, pp. 45–59.

Hirano, S. 1989a. "Some Commercial Products Made of Chitin and Chitosan," *MOL,* pp. 27–31.

Hirano, S. and Nagao, N. 1989b. "Effects of Chitosan, Pectic Acid, Lysozyme, and Chitinase on the Growth of Several Phytopathogens," *Agric. Biol. Chem.,* 53:3065–3066.

Hirano, S., Yamamoto, T., Hayashi, M., Nishida, T. and Inui, H. 1990. "Chitinase Activity in Seeds Coated with Chitosan Derivatives," *Agric. Biol. Chem.,* 54:2719–2720.

Hirano, S., Itakura, C., Seino, H., Akiyama, Y., Nonaka, I., Kanbara, N. and Kawakami, T. 1990a. "Chitosan as an Ingredient for Domestic Animal Feeds," *J. Agr. Food Chem.,* 38:1214–1217.

Hirano, S., Seino, H., Akiyama, Y. and Nonaka, I. 1990b. "Chitosan: A Biocompatible Material for Oral and Intravenous Administrations," in *Progress in Biomedical Polymers,* Gebelein, C. G. and Dunn, R. L., eds. New York: Plenum Press, pp. 283–290.

Hirano, S., Koishibara, Y., Kitaura, S., Taneko, T., Tsuchida, H., Murae, K. and Yamamoto, T. 1991a. "Chitin Biodegradation in Sand Dunes," *Biochem. System. Ecol.,* 19:379–384.

Hirano, S., Iwata, M., Yamanaka, K., Tanaka, H., Toda, T. and Inui, H. 1991b. "Enhancement of Serum Lysozyme Activity by Injecting a Mixture of Chitosan Oligosaccharides Intravenously in Rabbits," *Agric. Biol. Chem.,* 55:2623–2625.

Hirano, S. 1993. "Molecular Mechanism of the Conservation of the Eco-system by the Circulation of Chitinous Compounds on the Earth," (in Japanese), *Res. Environ. Earth,* 27:81–113.

Hirano, S., Inui, H., Kosaki, H., Uno, Y. and Toda, T. 1994a. "Chitin and Chitosans: Ecologically Bioactive Polymers," in *Biotechnology and Bioactive*

Polymers, Gebelein, C. and Carraher, C., eds. New York: Plenum Press, pp. 43–54.

Hirano, S. and Zhang, M. 1994b. "Hydrogels of *N*-Substituted Derivatives of Chitosan, and Some of Their Applications," in *Proceedings of 3rd International Marine Biotechnology Conferences,* August 7–12, 1994, Tromso, Norway.

Hirano, H., Inui, H. and Yamamoto, K. 1994c. "The Mineralization of CO_3^{2-} Ions in Crab Shells, and Their Mimetic Composite Materials," in *Proceedings of 2nd International Conference on Carbon Dioxide Removal,* Oct. 24–27, 1994, Kyoto, p. 136.

Hirano, S., Yamada, M., Yamamoto, K., Inui, H. and Ji, M. 1995. "Chemical Mineralization of CO_3^{2-} Ions on the Surface of a Chitosan-CuCl$_2$ Chelate Membrane," *1st International Conference of the European Chitin Society,* Sept. 11–13, Brest.

Hirano, S. and Akiyama, Y. 1995. "Absence of a Hypocholesterolemic Action in High-Serum-Cholesterol Rabbits," *J. Sc. Food Agric.,* 69:91–94.

Hutadilok, N., Mochimasu, T., Hisamori, H., Hayashi, K., Tachibana, H., Ishii, T. and Hirano, S. 1995. "The Effect of *N*-Substitution on the Hydrolysis of Chitosan by an Endochitosanase," *Carbohydr. Res.,* 268:143–149.

Iida, J., Une, T., Ishihara, C., Nishimura, K., Tokura, S., Mizukoshi, N. and Azuma, I. 1987. "Stimulation of Non-specific Host Resistance against Sendai Virus and *E. coli* Infections by Chitin Derivatives in Mice," *Vaccine,* 5:270–274.

Inui, H., Kosaki, H., Uno, Y., Tabata, K. and Hirano, S. 1991. "Induction of Chitinases in Rice Callus Treated with Chitin Derivatives," *Agric. Biol. Chem.,* 55:3107–3109.

Inui, H., Tsujikubo, M. and Hirano, S. 1995. "Low Molecular Weight Chitosan Stimulates Mitogenic Response to Platelet-derived Growth Factor in Vascular Smooth Muscle Cells," *Biosci. Biotechnol. Biochem.,* 59:2111–2114.

Japanese Soc. on Chitin/Chitosan, ed. 1988. *Chitin and Chitosan* (in Japanese), Tokyo: Gihodo Publishing Co., Inc.

Jap. Soc. Chitin/Chitosan, ed. 1990. *Applications of Chitin and Chitosan* (in Japanese). Tokyo: Gihodo Publishing Co., Inc.

Jap. Soc. Chitin/Chitosan, ed. 1995. *A Handbook of Chitin and Chitosan* (in Japanese). Tokyo: Gihodo Publishing Co., Inc.

Kifune, K. 1995. "Biomedical Materials," (in Japanese), in *A Handbook of Chitin and Chitosan,* Jap. Soc. Chitin/Chitosan, ed. Tokyo: Gihodo Publishing Co., pp. 323–354.

Kikuchi, Y. 1975. "Polyelectrolyte Complex of Heparin with Chitosan," *Makromol. Chem.,* 175:2209–2211.

Knorr, D. 1984. "Use of Chitinous Polymers in Food–A Challenge for Food Research and Development," *Food Technol.,* 38:85–89.

Kobayashi, T., Otsuka, S. and Yugai, Y. 1979. "Effect of Chitosan on Serum and Liver Cholesterol Levels in Cholesterol-fed Rats," *Nutr. Rep. Int.,* 19:327–334.

Lorito, M., Broadway, R. M., Hayes, C. K., Woo, S. L., Noviello, C., Williams, D. L. and Harman, G. E. 1994. "Proteinase Inhibitors from Plant as a Novel Class of Fungicides," *Mol. Plant-Microbe Interact.,* 7:525–527.

Miyazaki, S., Yamaguchi, H., Yokouchi, C., Takada, M., Hou, W. M. 1988. "Sustained Release of Indomethacin from Chitosan Granules in Beagle Dogs," *J. Pharm. Pharmacol.,* 40:642.

Miyazaki, S., Yamaguchi, H., Takada, M., Hou, W. M., Takeichi, Y. and Yasubuchi, H. 1990. "Pharmaceutical Applications of Biomedical Polymers. 29. Preliminary Study on a Film Dosage Form Prepared from Chitosan for Oral Drug Delivery," *Acta Pharm. Nord.,* 2:401–406.

Muzzarelli, R. A. A., Roccheti, R. and Marangio, G. 1972. "Separation of Zirconium, Niobium, Cesium and Ruthenium for the Determination of Cesium in Nuclear Fuel Solutions," *J. Radioanal. Chem.,* 10:17–26.

Muzzarelli, R. A. A. 1973. *Chitin,* Oxford: Pergamon Press.

Muzzarelii, R. A. A., Mattiolibelmonte, M., Tietz, C., Biagini, R., Ferioli, G., Brunelli, M. A., Fini, M., Giardino, R., Ilari, P. and Biagini, G. 1994. "Stimulatory Effect on Bone Formation Exerted by a Modified Chitosan," *Biomaterials,* 15:1075–1081.

Nagyvary, J. J., Falk, J. D., Hill, M. L., Schmidt, M. L., Wilkins, A. and Brodburg, E. L. 1979. "The Hypolipidemic Activity of Chitosan and Other Polysaccharides in Rats," *Nutr. Rep. Int.,* 20:677–684.

Nihon Kogyokai, ed. 1987. *The Development of Applications of Chitin and Chitosan* (in Japanese), Tokyo: Nihon Kogyokai, Inc.

Nishimura, K., Nishimura, S., Nishi, N., Saiki, I., Tokura, S., and Azuma, I. 1984. "Immunological Activity of Chitin and Its Derivatives," *Vaccine,* 3:379–384.

Nishimura, K., Nishimura, S., Nishi, N., Numata, F., Tone, Y., Tokura, S. and Azuma, I. 1985. "Adjuvant Activity of Chitin Derivatives in Mice and Guinea Pigs," *Vaccine,* 3:379–384.

Nishimura, K., Ishihara, C., Ueki, S., Tokura, S. and Azuma, I. 1986. "Stimulation of Cytokine Production in Mice Using Deacetylated Chitin," *Vaccine,* 4:151–156.

Nishimura, K., Nishimura, S., Seno, H., Nishi, N., Tokura, S. and Azuma, I. 1987. "Macrophage Activation with Multiporous Beads Prepared from Partially Deacetylated Chitin," *Vaccine,* 5:136–140.

Notsu, S., Saito, N., Kosaki, H., Inui, H. and Hirano, S. 1994. "Stimulation of Phenylalanine Ammonia-lyase and Lignification in the Rice Callus Treated with Chitin, Chitosan, and Their Derivatives," *Biosci. Biotechnol. Biochem.,* 58:552–553.

Ohtakara, A. 1995. "Enzymatic Degradation of Chitin and Chitosan," in *A Handbook of Chitin and Chitosan,* Jap. Soc. Chitin/Chitosan, ed. Tokyo: Gihodo, Co., Inc., pp. 54–86.

Sakaguchi, T., Horikoshi, T. and Nakajima, A. 1979. "Adsorption of Uranium from Sea Water by Biological Substances," *Nippon Nogei Kagaku Kaishi,* 53:211–217.

Sano, H., Shibasaki, T., Itoi, H. and Takaesu, Y. 1991. *Bull. Tokyo Dent. Coll.,* 32:9–17.

Schlumbaum, A., Mauch, F., Vogeli, U. and Boller, T. 1986. "Plant Chitinases Are Potent Inhibitors of Fungal Growth," *Nature,* 324:365–367.

Shibasaki, K., Sano, H., Matsukubo, T. and Takaesu, Y. 1994. "pH Response of Human Dental Plaque to Chewing Gum Supplemented with Low Molecular Chitosan," *Bull. Tokyo Dent. Coll.,* 35:27–32, 33–39, 61–66.

Shibasaki, K., Matsukubo, T. and Takaesu, Y. 1995. "Cariostatic Effect of Low Molecular Weight Chitosan in Rats," *Shika Gakuho,* 95:1–17.

Shinoda, K. and Nakajima, A. 1975. "Complex Formation of Heparin Sulfated Cellulose, Hyaluronic Acid or Chondroitin Sulfate," *Bull. Inst. Chem. Res. Kyoto Univ.,* 53:392–399, 400–408.

Song, Y., Onishi, H., and Nagai, T. 1993. "Conjugate of Mitomycin C with *N*-Succinylchitosan: in vitro Drug Release Properties, Toxicity and Antitumor Activity," *Int. J. Pharm.,* 98:121–130.

Sugano, M., Fujikawa, T., Hiratsuji, Y. and Hasegawa, Y. 1978. "Hypocholesterolemic Effects of Chitosan in Cholesterol-fed Rats," *Nutr. Rep. Int.,* 18:531–537.

Sugano, M., Fujikawa, T., Hiratsuji, Y., Nakashima, K., Fukuda, N. and Hasegawa, Y. 1980. "A Novel Use of Chitosan as a Hypocholesterolemic Agent in Rats," *Am. J. Clin. Nutr.,* 33:787–793.

Sugano, M., Watanabe, S., Kishi, A., Izume, M. and Ohtakara, A. 1988. "Hypocholesterolemic Action of Chitosan with Different Viscosity in Rats," *Lipids,* 23:187–191.

Suzuki, K., Mikami, T., Okawa, Y., Tokoro, A., Suzuki, S. and Suzuki, M. 1986. "Antitumor Effect of *N*-Hexaacetylchitohexaose and Chitohexaose," *Carbohydr. Res.,* 151:403–408.

Tokoro, A., Kobayashi, M., Tatewaki, N., Suzuki, K., Okawa, Y., Mikami, T., Suzuki, S. and Suzuki, M. 1989. "Protective Effect of *N*-Acetylchitohexaose on *Listeria monocytogenes* Infection in Mice," *Microbiol. Immunol.,* 33:357–367.

Tsugita, T. 1995. "Agricultural Farming Materials," in *A Handbook of Chitin and Chitosan,* Jap. Soc. Chitin/Chitosan, ed. Tokyo: Gihodo, Co., Inc., pp. 440–458.

Uchida, Y. 1988. "Antibacterial Function," in *A Handbook of Chitin and Chitosan,* Jap. Soc. Chitin/Chitosan, ed. Tokyo: Gihodo, Co., Inc., pp. 302–321.

Vahouny, G. V., Satchithanadam, S., Cassidy, M. M., Lishtfort, F. B. and Furda, I. 1983. "Comparative Effects of Chitosan and Cholestyramine on Lymphatic Absorption of Lipids in the Rat," *Am. J. Clin. Nutr.,* 38:278–284.

Yamada, A., Shibuya, N., Kodama, O. and Akatsuka, T. 1993. "Induction of Phytoalexin Formation in Suspension-Cultured Rice Cells by *N*-Acetylchitooligosaccharides," *Biosci. Biotechnol. Biochem.,* 57:405–409.

PART II

STRUCTURE AND PROPERTIES

CHAPTER 3

Chitin Structure and Activity of Chitin-Specific Enzymes

MARIA L. BADE

Center for Cancer Research
Massachusetts Institute of Technology
Cambridge, MA 02139

Kinetically valid measurements of chitinases have been difficult to obtain, because of lack of a substrate that would give valid data with specific chitinases, as opposed to "chitinase activity," which also includes susceptibility to enzymes, such as lysozymes, experiments with "soluble chitin," "colloidal chitin," or totally nonspecific "chitinase" tests such as measuring rate of color production using "chitin azure" as substrate [1].[1] The search in my laboratory for a "good" chitin substrate, i.e., one that would give rapid product generation that is linear with time, was rewarded by successful isolation of Linear Chitin from animal exoskeletons. This is discussed in Chapter 15 of this book. In Linear form, both the native ultrastructure and the spacing between chitin assemblies are recaptured. Chitinases arising cyclically in the molt of arthropods or that are produced adaptively by microorganisms for their nourishment when they encounter chitin in the environment are programmed for a solid substrate. Therefore, it was no surprise that a substrate consisting of highly purified orderly chitin should be a solid. Insoluble but highly ordered Linear Chitin was developed, which gives

---

[1]The confusion in substrates has confused the enzyme terminology. EC 3.2.1.14 *chitinase* describes "random hydrolysis," which sounds like a description of endochitinase, whereas EC 3.2.1.52 is a catch-all *β-N-hexosaminidase* "hydrolyzing a terminal . . . residue," i.e., subsumes chitobiase. Neither describes exochitinase activity, the enzyme for which is a separate entity and also has distinct mechanisms depending on origin and/or function.

acceptable and reproducible kinetic data with chitinases and other chitin-specific enzymes [2].

Using Linear Chitin in enzyme studies, separate assays for exo- and endochitinase activity were formulated. Rapid and linear generation of Elson-Morgan color, or rapid linear generation of counts from *in vivo* labeled Linear Chitin, signaled the presence of exochitinase. Since endochitinase would not generate soluble product as rapidly, diminution of light scattering was looked for as this enzyme rapidly scissions larger chitin particles into smaller ones. We were led to this view by an observation recorded by Lunt and Kent [3]. Results were obtained in keeping with the expectations described above. Figures 1a and 1b show some of these. Exoactivity gives a straight line for product generation, whereas measurement of endoactivity gives a concave curve consistent with what is observed with endocellulases [4]. Obtaining straight-line exoactivity permits calculation of enzyme parameters like the Michaelis-Menten constant K_M, K_{cat}, V_{max}, etc. (Bade, manuscript in preparation).

An early result observed with exochitinase activity showed that exoenzymes require access to free ends of chitin assemblies, irrespective of whether the enzymes are derived from procaryotic or eucaryotic sources [2]. When activity of several *Streptomyces* chitinases was examined, catalog No. C6137 (Sigma) was found to be pure exoactivity, i.e., no endoactivity was apparent, whereas No. C 1525 (Sigma) has predominantly endoactivity. Use of the former in a digestion of partially collapsed Linear Chitin resulted in formation of a visible residue; a photomicrograph is shown in Figure 2. Vigorous agitation of Linear Chitin with a component of very thin strands gives rise to a test substrate, shown in Figure 3, consisting of semirigid rods of intact Linear Chitin and (originally very thin) strands that have collapsed and appear amorphous. This test material permits an easy qualitative assay for exo- or endochitinase activity: if exhaustive, i.e., overnight, digestion leaves a residue visible to the naked eye, the activity is pure exo, whereas absence of a residue implies endoactivity.

With this qualitative assay, we were able to separate on a Maley-Tarentino column [5] *Streptomyces plicatus* chitinase activity generated in mass culture by organisms presented with chitin as the sole carbon and nitrogen source. Figure 4 shows the conditions and readings from this column. The endoenzymes come off the diethylaminoethyl (DEAE)-cellulose column with dilute buffer, whereas a shallow salt gradient is required to dislodge the exoenzymes [6]. A result of this separation was the additional observation that *Streptomyces* exochitinase(s) exhibit the high-temperature optimum above 50°C generally ascribed to "chitinases," whereas the endochitinase(s) have their optimum between

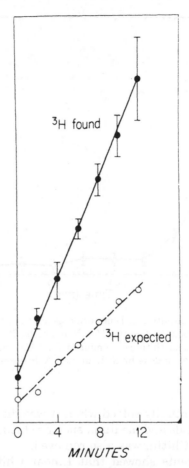

Figure 1a. Activity of exochitinase. Straight-line generation of product from *in vivo* labelled Linear Chitin, sulfate-stabilized, from *Manduca sexta* cuticle. Source of exochitinase: *Manduca* molting fluid. This panel is part of an experiment showing processivity of insect exochitinase acting on homologous chitin (reprinted from Reference [7] with permission from Academic Press, Inc.). Shown are average of two scintillation counts from duplicate experimental tubes ± error observed for each time point shown.

10° and 15°C; they are very active near 0°C and totally inactive above 40°C.

Another observation made in the course of this work was the following. To find chitinases of any type in the culture medium, collapsed chitin (i.e., chitin showing a high degree of disorder; cf. Figure 4 in Chaper 15 of this book) had to be employed in adapting the growing *Streptomyces* to a chitin diet. When Linear Chitin was employed for this purpose, no chitinase activity could be found in the culture medium.

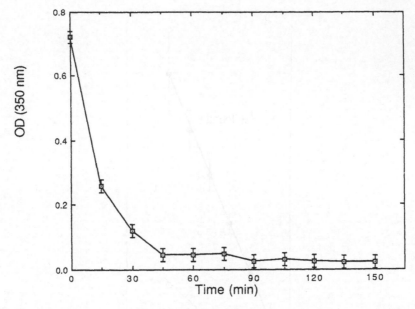

Figure 1b. Activity of endochitinase. Diminution of light scattering in a suspension of Linear Chitin particles. Source of chitin: *Manduca* cuticle, SO$_4$-stabilized. Source of enzyme: Partially purified endochitinase (combined Fr. 1-120, cf. Figure 4) from *Streptomyces plicatus*. Original work done by K. Laurence, A. Stinson, and K. Hickey.

This puzzle was eventually solved when it was discovered that during the separation of solids and medium, specific chitinase activity, tightly adsorbed to Linear Chitin, was also removed.

Separate experiments showed that Linear Chitin adsorbs specific chitinases quickly and tenaciously, as reported earlier by the authors and others. Figure 5 depicts a densitometer trace of a disc gel on which chitinase activity was separated. The origin of this (exo)enzyme was *Manduca* molting fluid allowed to adsorb to Linear Chitin; the inset shows a similar trace from whole molting fluid. The trace for chitinases, clearly very different from that for whole molting fluid, was obtained by adsorbing molting fluid enzymes to Linear Chitin and then washing it exhaustively. In addition to endo- and exochitinase activity, molting fluid contains much chitobiase. This activity rinses off nearly totally in the first two washings of Linear Chitin with enzymes specific for chitin adsorbed to it. Endochitinase activity washes out in two or three more rinses of the chitin, but extraction of exochitinase activity proved impossible. The only separation for Linear Chitin-adsorbed exochitinase came when sodium dodecyl sulfate (SDS) disc electrophoresis

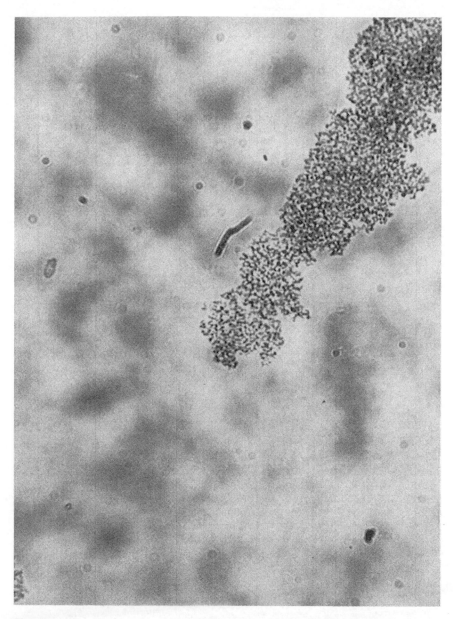

Figure 2. Chitin "clump." Residue of collapsed chitin after exhaustive digestion of Linear *Manduca* Chitin. Enzyme source: purified *Streptomyces* exochitinase. 400× magnification, transmission light microscopy.

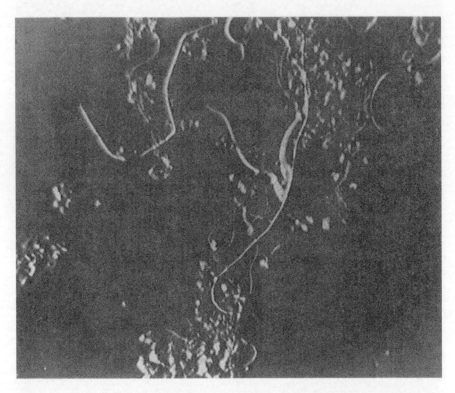

Figure 3. Linear Chitin test substrate. Linear Chitin after vigorous stirring with a magnetic bar of a suspension in phosphate-acetate buffer for 5 min at 37°C. Intact stabilized strands and collapsed, originally very thin, strands of chitin are apparent. Source of chitin: *Manduca* cuticle, SO$_4$-stabilized. Light microscopy, Hofmann Modulation Optics, 400× magnification.

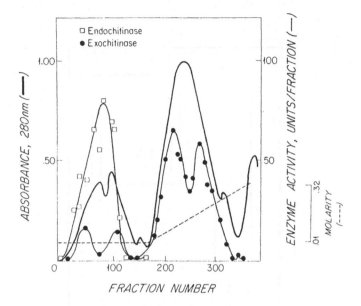

Figure 4. Separation of chitinases. Column effluent from a Maley-Tarentino column [5] of DEAE-cellulose charged with saturated $(NH_4)_2SO_4$-precipitated chitinases from a *Streptomyces plicatus* culture growing on chitin. 1 ml fractions were collected; every ten fractions combined and tested with test substrate like that shown in Figure 3. Protein by continuous flow through a UV-detector. Calculated molarity.

was used on adsorbed exochitinase activity. Molting fluid exochitinase activity requires calcium ions for activity and for adsorption [7]. When ethyleneglycol-bis-(β-aminoethyl ether) N,N,N',N'-tetraacetic acid (EGTA) was added to the reaction mixture after exoactivity had been activated through addition of calcium ion, activity was lowered by only 4%.

We tested whether exochitinase activity from molting fluid was more or less vulnerable to attack by a protease after it had become attached to Linear Chitin. It proved to be partially protected from proteolytic action but became much more vulnerable to destruction by freezing (Bade, manuscript in preparation). Molting fluid exoactivity was also found to be processive [8].

A log-log conversion of velocity versus substrate concentration for exochitinase derived from molting fluid was compared with one constructed with similar data obtained for a *Streptomyces*-derived exochitinase. Results are given in a Hill plot (see Figure 6). It is apparent that a *Streptomyces*-derived exochitinase gives the slope of 1 expected for a simple hydrolase, whereas the developmental insect molting fluid

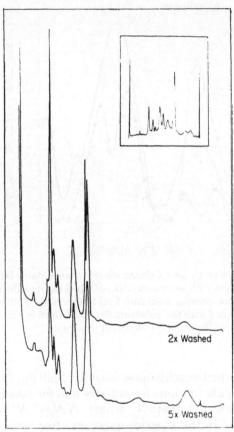

Densitometer traces.
Proteins adsorbed to washed chitin.

Figure 5. Densitometer trace of insect exochitinase. Chitinase-attacking enzymes from *Manduca* molting fluid were allowed to adsorb to Linear Chitin (0.1-ml molting fluid incubated for 15 min at 25°C with 12 mg Linear Chitin) and washed as shown with buffer (phosphate-acetate, 0.1 M, pH 6.5). "Chitin-chitinase" samples were then incubated for 2 h with 1% SDS plus 0.1% mercaptoethanol at 37°C. 10 μg protein per gel was marked with bromphenol blue, layered on a 7.5% SDS-polyacrylamide gel, and proteins remaining adsorbed to the chitin were separated in 0.1 M phosphate buffer in a field of 8 mA/gel. Gel columns were stained with Coomassie Blue. A trace of whole molting fluid from the same sample and separated under the same conditions is shown in the inset.

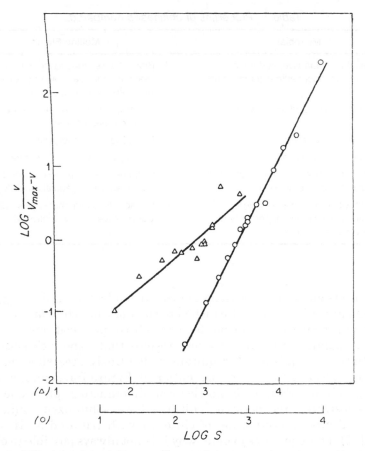

Figure 6. Hill plots of exochitinases. Kinetic data from twenty-nine separate experiments obtained with *Manduca* Linear Chitin were recalculated and plotted as shown: Δ-*Streptomyces* exochitinase; ● — *Manduca* exochitinase. (Note different abscissa for the two curves; this was done to avoid confusion through overlap of the two curves.)

chitinase gives a slope of $2+$ (1.98–2.37, with different enzyme and substrate preparations). These figures can be interpreted as denoting three strongly interacting sites or six weaker ones, but the symmetry is very good for three [9].

These data show that incisive studies on the mechanism of chitinases are now within reach, since the identical substrate was employed in studies of chitinases obtained from different biological sources and exhibiting different types of activity. Table 1 contrasts mechanisms for exochitinase activity derived from *Manduca* molting fluid and from *Streptomyces* culture medium. It must be emphasized here that exo-

Table 1. Properties of chitinases compared.

Microbial	Molting Fluid
1. Biosynthesis adaptive, system inducible only by accessible external chitin	1. Biosynthesis induced by hormone reaction; cuticular chitin not exposed until later in molting process
2. Separate endo- and exoactivities	2. Separate endo-, exo-, and N-hexosaminidase activities
3. Exo product: chitobiose	3. Exo product: N-acetylglucosamine
4. Exo attack pattern: random	4. Exo attack pattern: processive
5. Does not require Ca^{2+} ions, activation; does not exhibit regulation	5. Requires Ca^{2+} ions, enzymatic activation; calmodulin regulation likely
6. Exoactivity has simple Michaelis-Menten kinetics; no cooperativity of catalytic sites; no control	6. Exoactivity is oligomeric, positively cooperative, highly controlled

chitinase is virtually impossible to measure unless a reliable supply of free and accessible ends is furnished to the enzyme. This milieu is provided by Linear Chitin but not by other chitin preparations.

Close examination of preparations reveals that some lack necessary structural prerequisites. A requirement for freely accessible ends for evaluating exoactivity has been mentioned. Figure 7(a) depicts a photomicrograph and Figure 7(b) an electron photomicrograph of one often used substrate, "regenerated" chitin, where solubilized counts are measured after reacetylation of chitosan with tritiated acetic anhydride [10]. These pictures explain why it is not always possible to obtain straight-line data when exochitinase is employed with this material [11]. If activity is partially endo, free ends will be produced suitable for attack by exochitinase.

Another form, "microcrystalline" chitin, has been employed in various biotechnical applications [12,13]. A paper published in 1982 [14] showed a number of entities designated as microcrystals, but it was also shown that phosphate ester bonds claimed in the relevant patent could be removed by dialysis, and therefore, unlike what is observed in Linear Chitin, these ester functions do not form an integral part of the product. When microcrystalline chitin is examined in the polarizing microscope, some crystalline material can be discerned, but the bulk of the chitin mass is clearly collapsed (Figure 8; cf. Figure 4 in Chapter 15 of this book). On the other hand, Japanese commercial chitin can, in the polarizing microscope, be seen to be identical to the Compacted

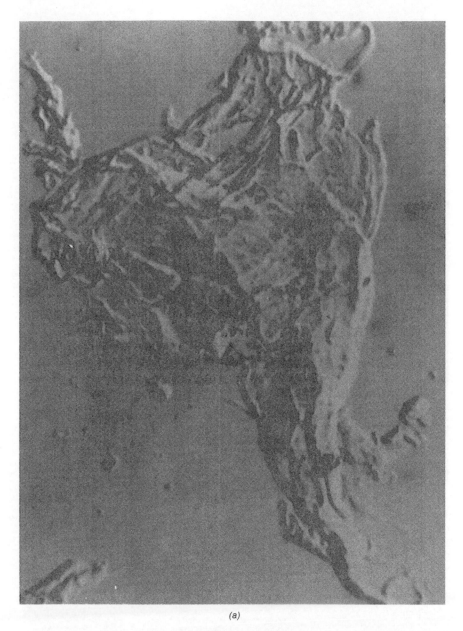

(a)

Figure 7. Microscopy of reconstituted "chitin." (a) Light microscopy of chemically altered chitin. Chitosan reacetylated with ³H-acetic anhydride [10]. 400× magnification, Hofmann Modulation Optics. (b) Scanning electron microscopy of chemically altered chitin, same material as (a). 1,000× magnification.

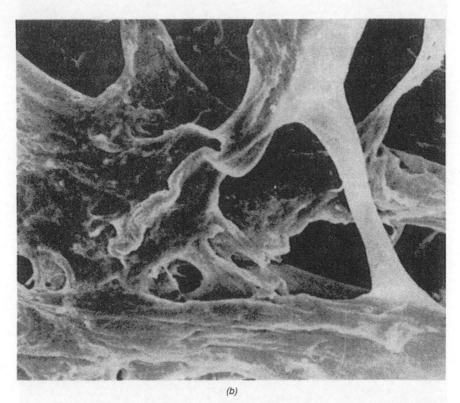

(b)

Figure 7 (continued). Microscopy of reconstituted "chitin." (a) Light microscopy of chemically altered chitin. Chitosan reacetylated with ^{3}H-acetic anhydride [10]. 400× magnification, Hofmann Modulation Optics. (b) Scanning electron microscopy of chemically altered chitin, same material as (a). 1,000× magnification.

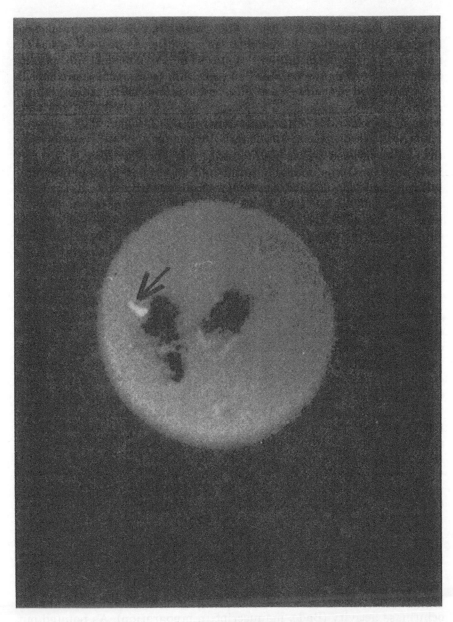

Figure 8. Microcrystalline chitin. Chitin made according to Reference [13]. Photographed in plane polarized light at 250× magnification. See arrow for green fluorescent microcrystal; the black clumps are amorphous chitin, which are part of the same preparation.

Chitin which is an intermediate in the making of Linear Chitin (U.S. Pat. No. 4,598,011 to Bade). The Japanese form retains the molecular fine structure without recapturing spacing between chitin assemblies (cf. Figures 8(b) and 8(c) in Chapter 15 of this book). Our studies on insect molting fluid chitinase, which generates single N-acetylglucosamine product [8] rather than chitobiose as is common with chitinases of procaryotic origin, eventually resulted in an insect chitinase model that accounts for both endo- and exochitinase activity from molting fluid [9]. In sum, the argument goes as follows: Because of multiple hydrogen bonds in secondary chitin structure, i.e., within microfibrils, it is not possible for endochitinase to insert the H_2O bonds required in the hydrolysis of glycosidic bonds in the interior of chitin chains because sideways slippage of a scissioned chain is prevented. If, however, the Rees and Scott model for structure of individual chitin chains is accepted, then the bonds involved in the phi-psi angle in the primary structure are strained [15]: 141° for the phi/psi angle versus the normal tetrahedral angle of 109.5°. A thermodynamic force driving toward the nonstrained tetrahedral angle for C1′-linked to O and thus wrenching a short length of the primary chain being cut, out of horizontal alignment with the others to which the sugar molecule is bonded, might drive separation of a sugar ring scissioned at the glycosidic linkage so that OH⁻ and H⁺ have room to bond to newly freed ends within the chitin chain. An added driving force that can be expected to make the separation of the short length from adjacent chains irreversible is that hydrogen bonding to H_2O molecules will replace the stabilizing hydrogen bonding to neighboring chains in the microfibril. Endochitinase seems to become loosely bound where it attacks chitin chains so that novel free ends form, and it seems reasonable to expect that the enzyme will assist the separation of one or a few residues to permit bonding of the elements of a water molecule to newly separated C1 and C4′ ends. Openings thus created will then permit the exochitinase from molting fluid to become inserted between adjacent chains and, as it processes, cooperatively scissioning three adjacent chains at the same time. This is in part confirmed by the above-mentioned near impossibility of rinsing off the exochitinase compared with chitobiase or endochitinase, as well as by the fact that, on a sugar gradient, insect exochitinase tends to form a dimer during high-speed centrifugation that still exhibits exochitinase activity (Bade, manuscript in preparation). As pointed out below, it is also compatible with the slope of insect exochitinase on the Hill plot. The mechanism proposed would account for many observations: processivity, cooperative behavior, and the trace obtained after exhaustive washing where structure of the exochitinase from molting fluid is clearly seen to be oligomeric. It is also consistent with partial protection of the exoactivity from protease digestion and its known vulnerability to freezing when associated with chitin.

The proposed tripartite symmetry is of particular interest. Figure 9 depicts a diagram showing the proposed chitinase model. Although tripartite symmetry in enzyme structures is as rare today as it is in the symmetry of organisms now extant, a number of seemingly tripartite fossil forms have been recovered, e.g., from the Ediacarian fauna [16]. The organisms in that formation were living in the Late Proterozoic, i.e., before plants began to colonize the land. Kelps, diatoms, etc., have structural and storage carbohydrates other than cellulose. Invasion of dry land was carried out by green algae that contained cellulose, which is employed to form the bulk of supporting tissues in terrestrial plants. But chitinous carapaces, e.g., in the form of trilobites and other invertebrate marine organisms, clearly existed in large numbers millions of years before terrestrial plants grew cellulose-supported roots, stems, leaves, and flowers. It seems likely, therefore, that large-scale biosynthesis of chitinases preceded large-scale employment of cellulases in geological time. Thus, tripartite symmetry in chitinase systems might be a surviving relic.

Strong homologies between chitinase and cellulase systems appear to exist. Chitinase from the yeast species *Saccharomyces cerevisiae* shows some sequence identity to chitinases from a variety of microorganisms, endo-H enzyme from *S. plicatus*, and a yeast killer toxin from a

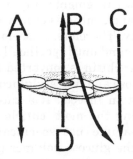

Figure 9. Model of insect exochitinase. Explanation: A, B, and C are chitin chains directed as indicated by the arrows. Three exochitinase catalytic units (round discs) are working cooperatively to free single *N*-acetylglucosamine units; other experiments showed that the endochitinase from molting fluid tends to cut over time through a number of chains in the same location on the strand (Bade, manuscript in preparation). The striped enzyme unit is the "anchor site" required as part of a processive oligomeric enzyme. The chain along which it moves is shown as being anti-parallel, but the argument can accommodate a parallel direction for chain D as well. Other proteins are also involved (shown stippled); one of these, a poorly staining protein running very fast in the gel, may be calmodulin (T. Vanaman, personal communication).

Kluyveromyces species. According to its sequence, yeast chitinase can be divided into four domains: a signal sequence, a catalytic region, a serine-threonine-rich linker region, and the chitin-binding domain. The last, although not enhancing the rate of chitin hydrolysis, is required for localization of the enzyme on its substrate and bears a resemblance to noncatalytic high-affinity binding domains for their substrates that recently were found in a number of cellulose- and starch-degrading enzymes. Further, the chitinase binding domain from *S. cerevisiae* has two regions where the sequence is identical with sequences in the cellulose-binding peptide from *Trichoderma reesei* [17]. It appears probable, therefore, that an ur-enzyme for splitting glycosidic bonds in carbohydrates evolved into a family of specific enzymes for the hydrolytic scissioning of chitin, cellulose, and assorted glucans. A necessity for binding would also be consistent with requirements for attack on the interior of chitin or cellulose chains in endoactivity where, as explained above, it is necessary to break some of the tight hydrogen bonds that bind individual sugars to sugars in neighboring chains.

The complexity of insect molting fluid exochitinase, shown in Figure 5, might be invoked to explain another puzzling finding for eucaryotic molting fluid chitinase activity. It was shown in 1962 for the molt in insects [18], later confirmed for crustacea [19], that specific activity of carbon in chitin is the same in old cuticle being shed as it is in the new cuticle being laid down. Yet, no soluble sugar fraction in molting fluid through which breakdown fragments might be expected to pass between breakdown of old and synthesis of new cuticle, has the requisite high carbon radioactivity. Therefore, it seems as if chitin fragments from old cuticle remain bound until specifically transferred to new cuticle. Any surplus cuticular chitin is converted to glucose and laid down as glycogen. Indeed, glucose and *N*-acetylglucosamine seem to be freely interconvertible, at least in insects. If a caterpillar is injected with radioactive glucose during the molt, cuticle chitin becomes labeled; likewise, *N*-acetylglucosamine not specifically required for cuticle resynthesis appears as the glucose polymer glycogen, which has glycosidic linkages dissimilar to chitin. Caterpillars, which live to feed, preserve the material assimilated in each instar, and their stores are preserved as far as possible into the pupal and adult stages. The rapid disappearance of fat stores during spinning of the cocoon [20] is explained, not by fat-to-carbohydrate conversion during the larval → pupal molt, but by the animal's use of fat stores for energy production. Crustaceans have the same problem of macromolecule preservation, and in their case, it extends even to preservation of the calcium packing the integument to make it hard and capable of providing an origin attachment for their muscles. During the premolt stage, calcium salts are withdrawn from the integument and stored in long and closely packed crystalline columns in the anterior region of the cephalothorax, and when lobsters are prevented from ingesting their shed carapace, recalcification of the new and still soft carapace is delayed (Oak Bluff Lobster Hatchery, Martha's Vineyard, MA, personal communication). The reported conservation of the carbon in chitinous material in crustaceans, therefore, makes perfect sense, but the mechanism for this in either insects or crustaceans still lacks a biochemical explanation.

Chitinases from microorganisms demonstrably are simple hydrolases. Nevertheless, they too require prior attachment to the substrate as shown in Figure 10, which depicts the margin of a *Rhizopus* sp. yeast growing on Linear Chitin prepared from insect cuticles. For purposes of nutrition, the yeast adaptively produces enzymes that will allow degradation of chitin or chitosan encountered in the environment. The capacity for producing chitinases and chitosanases by microorganisms upon encountering chitin and/or chitosan was demonstrated by Eveleigh and co-workers for organisms in "any soil or sediment examined" [21]. This attachment is also implicit in the conserved carbohydrate-binding region common to chitinases, cellulases, and glucanases that was mentioned previously. Another enzyme acting on chitin is the (intracellular) deacetylase first reported by Ito and Akari [22]. When cell extract from *Rhizopus pseudochinensis*, which like all other *Mucorales* makes chitosan for its cell wall by deacetylating chitin, was allowed to act on doubly labeled Linear Chitin, deacetylation activity ceased entirely after about 30% of acetyl groups had been removed. It might be expected that specific deacetylase shares a binding region with chitinases and cellulases, but that does not appear to be the case. It was, however, found that the chelator, *o*-phenanthroline, completely inhibited the deacetylase activity in the extract. This phenomenon implies that a transition metal, e.g., zinc, is required for deacetylase activity (although there is a disclaimer in Reference [23] where deacetylation to 70% acetylation is also described). Further experiments with this enzyme seem very desirable. It is hoped that employment of pure deacetylase will result in chitosan with a reproducible pattern of deacetylation, since Fenton et al. [21] also reported that chitosanases act on chitin that is between 95% and 65% acetylated and are inactive outside this region of chitin acetylation.

CONCLUSIONS

1. Chitin-specific enzymes need substrates of known composition and structure for proper estimation of their activity and for comparisons

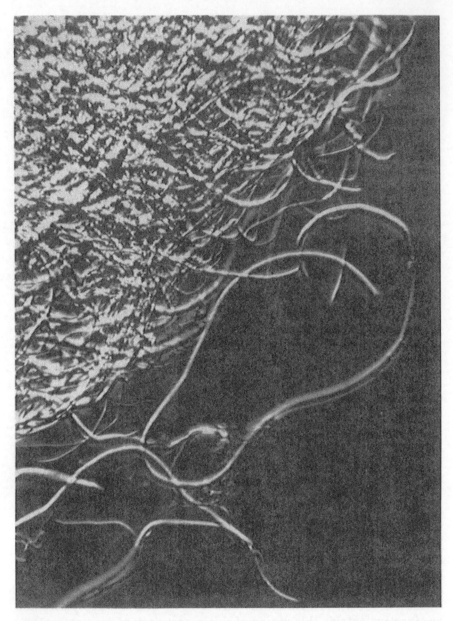

Figure 10. Insect Linear Chitin stimulating adaptive enzyme production. Shown is the margin of a small fungus (probably *Rhizopus pseudochinensis*); which is growing in a sample of *Manduca* Linear Chitin. It was fished out, put under a coverglass, squashed flat, and photographed at 400× magnification. Hofmann Modulation Optics.

between enzymes of varied biological activities. It is therefore necessary to inquire into the history of any chitin purification if generalizations are to be based on results obtained.

2. When a substrate is employed in enzyme studies, kinetics of endo- and exochitinase can be examined, and study of deacetylase is simplified because the amount of substrate acted upon can be unequivocally established.

3. When defined solid substrate is employed for these studies, endo- and exochitinases can be distinguished, and chitinases of different origin can be shown to have different mechanisms.

4. Accumulating evidence implies a common origin for carbohydrases, as well as similar organization, for chitinases, cellulases, and glucanases, in procaryotes and eucaryotes. Yeast chitinase has been shown to be divided into domains: a signal sequence (in the precursor enzyme), a catalytic region (which for insect exochitinase may exist in triplicate in each oligomeric enzyme), a serine-threonine-rich linker region, and a sequence required for attachment to the substrate.

ACKNOWLEDGEMENTS

Previously, this work was supported by National Institutes of Health (NIH) and National Science Foundation (NSF), before work on environmental matters became unfashionable. Grateful acknowledgement is made to my co-workers: Alfred Stinson, Senior Research Associate; Michael Kosmo, J. J. Shoukimas, Ivan Boyer, Kerry Hickey, and L. Lapierre, students; Nehad A.-M. Moneam, postdoctoral associate; and G. R. Wyatt, Ph.D. thesis advisor.

REFERENCES

1. Winicur, S. and H. K. Mitchell. 1974. "Chitinase Activity during *Drosophila* Development." *J. Insect Physiol.* 20:1795–1803; Hirano, S. 1988. "Water-Soluble Glycol Chitin and Carboxymethyl Chitin." Sect. [45] in *Methods in Enzymology Vol. 161: Biomass Pt. B, Lignin, Pectin, and Chitin.* pp. 408–410 ff. Also: Sections to [50] *ibid.* (W. A. Wood and S. T. Kellogg, eds.). San Diego, CA: Acad. Press. Also: Turner, C. D., D. Koga and K. J. Kramer. 1981. *Insect Biochem.* 11:215–219. Also: Sigma Catalogue #C 3020. Also: Hackman, L. H. and M. Goldberg. 1964. "New Substrates for Use with Chitinases." *Anal. Biochem.* 8:397–401.

2. Bade, M. L., A. Stinson and Nehad A.-M. Moneam. 1988. "Chitin Structure and Chitinase Activity: Isolation of Structurally Intact Chitins." *Conn. Tissue Res.* 17:137–151.

3. Lunt, M. R. and P. W. Kent. 1960. "Chitinase System from *Carcinus maenas.*" *Biochem. Biophys. Acta* 44:371–373.

4. Enari, T.-M. and M.-L. Niku-Paarola. 1988. "Nephelometric and Turbidimetric Assay for Cellulase." Sect. [11] in *Meth. Enzymol. 160: Biomass Pt. A: Cellulose and Hemicellulose.* pp. 117–126. San Diego: Acad. Press.

5. Tarentino, A. L. and F. Maley. 1974. "Purification and Properties of an Endo-β-N-Acetyl-glycosaminidase from *Streptomyces griseus.*" *J. Biol. Chem.* 249:811–817.

6. Bade, M. L. and K. Hickey. 1988. "Classification of Enzymes Hydrolyzing Chitin." In: *Proceed. 4th Internat. Confer. Chitin and Chitosan.* pp. 179–183. Published as: *Chitin and Chitosan* (Skjåk-BrAEk, G., T. Anthonsen and P. Sandford, eds.), London: Elsevier Appl. Sci.

7. Bade, M. L. and A. Stinson. 1981. "Biochemistry of Insect Differentiation. A System for Studying the Mechanism of Chitinase Activity in vitro." *Arch. Biochem. Biophys.* 206:213–221.

8. Bade, M. L. and A. Stinson. 1981. Biochemistry of Insect Differentiation. Requirement for High in vitro Moulting Fluid Chitinase Activity." *Insect Biochem.* 11:599–604.

9. Bade, M. L. and A. Stinson. 1981. "Chitin and Chitinase: A Kinetic Model." *J. Theor. Biol.* 93:697–700.

10. Molano, J., A. Duran and E. Cabib. 1977. "A Rapid and Sensitive Assay for Chitinase Using Tritiated Chitin." *Anal. Biochem.* 83:648–656.

11. Molano, J., I. Polacheck, A. Duran and E. Cabib. 1979. "An Endochitinase from Wheat Germ. Activity of Nascent and Preformed Chitin." *J. Biol. Chem.* 254:4901–4907.

12. Dunn, H. J., L. Farr and P. M. Farr. 1974. "Microcrystalline Chitin." U.S. Pat. No. 4,034,121.

13. Austin, P. R. and C. J. Brine. 1981. "Chitin Powder and Process for Making It." U.S. Pat. No. 4,286,087.

14. Deschamps, J. R. and J. E. Castle. 1982. "Microcrystalline Chitin: Its Preparation and Properties." In: *Chitin and Chitosan, Proc. 2nd Internat. Confer. Chitin/Chitosan.* pp. 63–65. Tottori: The Japanese Soc. of Chitin/Chitosan.

15. Rees, D. and W. E. Scott. 1971. "Polysaccharide Conformation. Part VI. Computer Model-Building for Linear and Branched Pyranoglycans. Correlation with Biological Function. Preliminary Assessment of Inter-Residue Forces in Aqueous Solution. Further Interpretation of Optical Rotation in Terms of Chain Conformation." *J. Chem. Soc.* (B): 469–480.

16. Wilson, E. O., T. Eisner, W. R. Briggs, R. E. Dickerson, R. L. Metzenberg, R. D. O'Brien, M. Susman and W. E. Boggs. 1973. "Picture of *Tribrachidium heraldicum.*" In *Life on Earth* (1st Ed.), p. 592. Stamford, CT: Sinauer Assoc.

17. Kuranda, M. J. and P. W. Robbins. 1991. "Chitinase is Required for Cell

Separation during Growth of *Saccharomyces cerevisiae. J. Biol. Chem.* 266:19758–19767.

18. Bade, M. L. and G. R. Wyatt. 1962. "Metabolic Conversions during Pupation of the *Cecropia* Silkworm. 1. Deposition and Utilization of Nutrient Reserves." *Biochem. J.* 83:470–478.

19. Hettick, P. P. and J. R. Stephenson. 1976. "Changing Activities of the Crustacean Epidermis during Incorporation of *N*-Acetylglucosamino-^{14}C and Glucose-^{3}H in Crayfish." *Am. Zool.* 16:237.

20. Couvreur, M. E. 1895. *Compt. Rend. Soc. Biol., Paris,* 47:796.

21. Fenton, D., B. Davis, C. Rotgers and D. E. Eveleigh. 1978. "Enzymatic Hydrolysis of Chitosan." In: *Proceed. 1st Internat. Confer. Chitin/Chitosan.* pp. 525–541. Cambridge, MA: MIT Sea Grant Program.

22. Akari, Y. and E. Ito. 1975. "Pathway of Chitosan Formation in *Mucor rouxii:* Enzymatic Deacetylation of Chitin." *Eur. J. Biochem.* 55:71–78.

23. Akari, Y. and E. Ito. (1988). "Assay for *Mucor* Deacetylase." Sect. [66] in *Meth. Enzymol. Vol. 161: Biomass Pt.B: Lignin, Pectin, and Chitin.*

β-Chitin and
Reactivity Characteristics

KEISUKE KURITA
Department of Industrial Chemistry
Faculty of Engineering
Seikei University
Musashino-shi, Tokyo 180, Japan

INTRODUCTION

Studies on chitin are progressing rapidly in various fields to seek potential applications based on the characteristic properties of this amino polysaccharide. They are, however, directed mostly to the ordinary chitin, α-chitin, owing primarily to the easy accessibility. There is another form of chitin, β-chitin. It is also distributed widely in nature, but squid pens may be the most promising source to isolate it in quantity. β-Chitin is interesting because of the unique properties different from those of α-chitin.

Since α-chitin is quite resistive to modification reactions because of the intractable nature due to the strong intermolecular hydrogen bonds, difficulty is generally encountered in achieving high degrees of substitution and also in preparing derivatives with well-defined structures. In sharp contrast, β-chitin shows considerable affinity for solvents (Austin et al., 1989) and is thus anticipated to show high reactivity compared with α-chitin. β-Chitin may thus become a promising candidate to conduct various reactions efficiently under mild conditions. This chapter describes some reactions of β-chitin and the derived chitosan to elucidate the high potential as versatile starting materials for modifications of chitin and chitosan.

79

ISOLATION OF β-CHITIN AND SOME PROPERTIES

β-Chitin is isolated from squid pens by treating with acid and alkali in a manner similar to that for α-chitin from shrimp or crab shells. Squid pens, however, swell in water; and, moreover, they are composed of chitin and proteins almost exclusively, with only trace amounts of metal salts such as calcium carbonate—unlike shrimp and crab shells. Milder conditions are thus adequate to isolate β-chitin (Kurita et al., 1993a).

Although chitin molecules are packed in an antiparallel fashion in α-chitin, they arrange in a parallel fashion in β-chitin (Gardner and Blackwell, 1975). This makes β-chitin show high affinity for organic solvents and water; it is soluble in formic acid and disperses well in water. As expected, it is much more hygroscopic and retains more water than α-chitin (Kurita et al., 1993a).

REACTIONS OF β-CHITIN

Deacetylation

Detailed studies on the deacetylation behavior of chitin are important to develop efficient procedures for preparing partially deacetylated chitins as well as chitosan:

$$\left[\begin{array}{c} \text{HO} \hspace{1.5cm} \text{OH} \\ \text{O} \\ \text{NHAc} \end{array} \right]_n \xrightarrow{\text{OH}^-} \left[\begin{array}{c} \text{HO} \hspace{1.5cm} \text{OH} \\ \text{O} \\ \text{NHAc} \end{array} \right]_l \left[\begin{array}{c} \text{HO} \hspace{1.5cm} \text{OH} \\ \text{O} \\ \text{NH}_2 \end{array} \right]_m \tag{1}$$

As shown in Figure 1, deacetylation of β-chitin proceeds much more facilely than that of α-chitin in aqueous sodium hydroxide (Kurita et al., 1993a). This is ascribable to the loose arrangement of the molecules with weak intermolecular hydrogen bonds. The reaction is efficient at a temperature as low as 60°C in 40% sodium hydroxide, unlike that of α-chitin. During the reaction, however, discoloration is observed and becomes evident with an increase in the reaction temperature. Deacetylation to high degrees should thus be carried out at 60–80°C to suppress discoloration.

N-Acetylation

Chitin is partially deacetylated in nature, and, furthermore, some acetyl groups are removed in the isolation process. Actually, chitin iso-

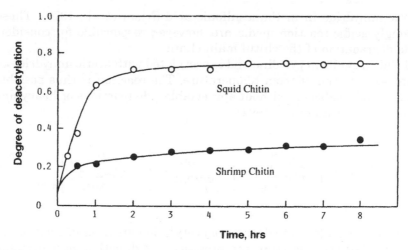

Figure 1. Deacetylation of α- and β-chitins in 30% sodium hydroxide at 100°C.

lated by a common procedure has a degree of deacetylation of 0.07–0.15. To discuss the structure-property relationship, fully *N*-acetylated chitin is necessary as a starting material to prepare derivatives with well-defined structures.

When β-chitin dispersed in methanol is subjected to the reaction with acetic anhydride, acetylation occurs selectively at the free amino groups. Full *N*-acetylation is attained at 40°C, and structurally uniform chitin, poly(*N*-acetylglucosamine), is obtained (Kurita et al., 1994):

$$
\left[\begin{array}{c} \text{OH} \\ \text{HO} \quad \text{O} \\ \text{NHAc} \end{array}\right]_l
\left[\begin{array}{c} \text{OH} \\ \text{HO} \quad \text{O} \\ \text{NH}_2 \end{array}\right]_m
\xrightarrow[\text{MeOH}]{\text{Ac}_2\text{O}}
\left[\begin{array}{c} \text{OH} \\ \text{HO} \quad \text{O} \\ \text{NHAc} \end{array}\right]_n
\tag{2}
$$

Acetylation of α-chitin, however, proceeds to only a limited extent. Selective *N*-acetylation also occurs with the water-soluble chitin (see Chapter 6 in this book) (Kurita et al., 1977).

Full Acetylation

Ordinary α-chitin is very resistive toward acetylation, and it is difficult to attain high degrees of acetylation. Full acetylation is accomplished only under rather harsh conditions, such as with acetic anhydride and hydrogen chloride (Schorigin and Hait, 1935) or with acetic anhydride in methanesulfonic acid (Nishi et al., 1979). These strongly acidic reaction media are, however, responsible for considerable degradation of the chitin main chain.

β-Chitin swells in pyridine and is acetylated with acetic anhydride to some extent even at room temperature. The reaction is thus possible under mild conditions without appreciable side reactions or heavy discoloration (Kurita et al., 1994):

(3)

As shown in Table 1, the degrees of acetylation are around 2 without a catalyst, but become 3.0 in the presence of 4-dimethylaminopyridine (DMAP) at 50°C. Under similar conditions, a degree of substitution of 1.8 is the highest value for α-chitin.

Tosylation

Chitin derivatives having p-toluenesulfonyl (tosyl) groups are versatile precursors for controlled further modifications, since they are soluble in organic solvents and highly reactive. Tosylation of α-chitin with tosyl chloride is, however, quite sluggish in pyridine but can be performed by interfacial condensation with an aqueous alkali chitin solution and a chloroform solution of tosyl chloride (see Chapter 6 in this book) (Kurita et al., 1991, 1992). During the reaction, partial deacetylation takes place because of the alkaline medium, and subsequent acetylation followed by O-deacetylation is necessary. These posttreatments are troublesome, and moreover, alkaline O-deacetylation is accompanied by a decrease in the degree of tosylation.

Table 1. Full acetylation of chitins.[a]

Chitin Source	(g)	Cat. (g)	Temperature (°C)	Time (h)	DS[b]
Squid	0.250	none	rt	72	2.07
Squid	0.209	Et$_3$N (10.7)	rt	48	1.95
Squid	0.211	DMAP (0.2)	rt	48	2.25
Squid	0.207	DMAP (0.2)	50	48	3.00
Shrimp	0.203	DMAP (0.2)	50	48	1.82

[a]In 20 ml of pyridine.
[b]Determined from the C/N ratio of elemental analysis.

Table 2. Tosylation of chitins.[a]

Chitin				
Source	(g)	DMAP (g)	Time (h)	DS[b]
Squid	0.204	0	24	0.34
Squid	0.202	0.2	24	0.62
Squid	0.203	0.2	48	0.70
Squid	0.200	1.0	48	0.80
Squid	0.201	2.0	48	0.83
Shrimp	0.200	0.2	48	0.18

[a]In 10 ml of pyridine at room temperature.
[b]Determined from the C/N ratio of elemental analysis.

If tosylation can be carried out in pyridine under mild conditions, no deacetylation takes place, and hence this alternative preparative procedure is more favorable for preparing tosyl chitins in a single step. With β-chitin, as expected, tosylation is successfully conducted in pyridine at room temperature in a swollen state (Kurita et al., 1994):

$$\left[\begin{array}{c} \text{OH} \\ \text{HO} \end{array} \begin{array}{c} \text{O} \\ \text{NHAc} \end{array} \text{O} \right]_n \xrightarrow[\text{Pyr}]{\text{TsCl}} \left[\begin{array}{c} \text{OTs} \\ \text{HO} \end{array} \begin{array}{c} \text{O} \\ \text{NHAc} \end{array} \text{O} \right]_l \left[\begin{array}{c} \text{OH} \\ \text{HO} \end{array} \begin{array}{c} \text{O} \\ \text{NHAc} \end{array} \text{O} \right]_m \qquad (4)$$

As summarized in Table 2, the degree of tosylation reaches above 0.8 easily. DMAP is again an effective catalyst. In contrast, the degree of tosylation of α-chitin is very low under these mild conditions.

Tritylation

The triphenylmethyl (trityl) group is typically used to protect primary hydroxyl groups, and 6-O-tritylated chitin is a key intermediate that enables discrimination of two kinds of hydroxyl groups of chitin for regioselective substitutions. Since α-chitin cannot be tritylated directly with trityl chloride, trityl chitin is prepared through five-step reactions: deacetylation of chitin, N-phthaloylation, tritylation, dephthaloylation, and final N-acetylation (Nishimura et al., 1990a, 1991b, 1993).

As in the other modifications, however, β-chitin allows direct tritylation in pyridine. Although no appreciable substitution is observed at 90°C without a catalyst, the reaction proceeds smoothly in the presence of DMAP:

$$\left[\begin{array}{c} \text{HO} \end{array} \begin{array}{c} \text{OH} \\ \text{O} \\ \text{NHAc} \end{array} \text{O}^- \right]_n \xrightarrow[\text{Pyr}]{\text{TrCl}} \left[\begin{array}{c} \text{HO} \end{array} \begin{array}{c} \text{OTr} \\ \text{O} \\ \text{NHAc} \end{array} \text{O}^- \right]_l \left[\begin{array}{c} \text{HO} \end{array} \begin{array}{c} \text{OH} \\ \text{O} \\ \text{NHAc} \end{array} \text{O}^- \right]_m \qquad (5)$$

The degree of tritylation reaches 0.75 under appropriate conditions (Table 3) (Kurita et al., 1994).

Acetolysis

Chitooligosaccharides and N-acetylchitooligosaccharides are becoming increasingly important in view of their interesting bioactivities including antitumor activity, antimicrobial activity, and plant growth regulation. They are also conveniently used for diagnostic determination of the activity of chitinolytic enzymes. The disaccharides, chitobiose and N,N'-diacetylchitobiose, are particularly useful as starting materials for tailored synthesis of model compounds for biologically significant glycoproteins and glycolipids (Nishimura et al., 1990b, 1990c, 1991a).

Acetolysis of α-chitin with acetic anhydride in concentrated sulfuric acid is one of the common methods for preparing a mixture of peracetylated chitooligosaccharides (Osawa, 1966). The yield of the disaccharide is generally less than 10% but can be raised to 16% when α-chitin is converted into colloidal chitin beforehand (Nishimura et al., 1989). β-Chitin goes into the solution much more easily, and the disaccharide is obtained in 17% yield without any special pretreatment (Kurita et al., 1994):

$$\left[\begin{array}{c} \text{HO} \end{array} \begin{array}{c} \text{OH} \\ \text{O} \\ \text{NHAc} \end{array} \text{O}^- \right]_n \xrightarrow[\text{H}_2\text{SO}_4]{\text{Ac}_2\text{O}} \text{AcO} \left(\begin{array}{c} \text{AcO} \end{array} \begin{array}{c} \text{OAc} \\ \text{O} \\ \text{NHAc} \end{array} \text{O} \right)_m \begin{array}{c} \text{OAc} \\ \text{O} \\ \text{NHAc} \end{array} \text{OAc} \qquad (6)$$

REACTION OF CHITOSAN DERIVED FROM β-CHITIN

As implied from the amorphous nature of chitosan derived from β-chitin (Kurita et al., 1993a), it may exhibit reactivity different from that of the ordinary chitosan derived from α-chitin. N-Phthaloylation (Kurita et al., 1982; Nishimura et al., 1990a) is thus examined to compare the reactivity characteristics of two types of chitosans. It is a convenient way to prepare a soluble precursor that makes possible regioselective modifications (see Chapter 6 in this book).

Table 3. Tritylation of chitins.[a]

Source	Chitin (g)	DMAP (g)	Time (h)	DS[b]
Squid	0.200	0	24	0
Squid	0.200	0.2	24	0.61
Squid	0.203	1.0	72	0.75
Shrimp	0.200	0.2	24	0.10

[a]In 10 ml of pyridine at 90°C.
[b]Determined from the C/N ratio of elemental analysis.

The reaction is carried out by heating suspended chitosan with phthalic anhydride in N,N-dimethylformamide:

(7)

As the substitution proceeds, a homogeneous solution results, and the reaction is complete in 5–7 h at 130°C with chitosan from α-chitin. With chitosan from β-chitin, however, the reaction is much more facile owing to its high affinity for the solvent, and heating for 1–2 h is sufficient (Table 4). It is noteworthy that the difference in the crystalline structures of chitins affects the reactivity even after deacetylation (Kurita et al., 1993b).

Table 4. N-Phthaloylation of chitosans.[a]

Source	Temperature (°C)	Time (h)	Appearance of Mixture	Yield (%)
Squid	100	24	homogeneous	85
Squid	130	1	homogeneous	88
Squid	130	3	homogeneous	91
Shrimp	100	48	heterogeneous	5
Shrimp	130	7	homogeneous	96

[a]Chitosan, 0.500 g; phthalic anhydride, 3 equiv; DMF, 20 ml.

CONCLUSION

The ordinary chitin, α-chitin, shows poor affinity for solvents, and thus chemical modifications generally encounter many difficulties, in particular, in preparing derivatives with well-defined structures and in attaining high substitution degrees. In sharp contrast, β-chitin is characterized by high affinity for both water and organic solvents, which enables various kinds of modification reactions to proceed efficiently under mild conditions in common solvents suitable for reactions. β-Chitin has thus been confirmed to be useful as a starting material for designing a sophisticated molecular environment leading to the development of advanced functions.

REFERENCES

Austin, P. R., J. E. Castle and C. J. Albisetti. 1989. "Beta-chitin from Squid: New Solvents and Plasticizers," in *Chitin and Chitosan.* G. Skjak-Braek, T. Anthonsen and P. Sandford, eds. Essex: Elsevier, p. 749.

Gardner, K. H. and J. Blackwell. 1975. "Refinement of the Structure of β-Chitin," *Biopolymers,* 14:1581.

Kurita, K., T. Sannan and Y. Iwakura. 1977. "Studies on Chitin. 3. Preparation of Pure Chitin, Poly(*N*-acetyl-D-glucosamine), from the Water-Soluble Chitin," *Makromol. Chem.,* 178:2595.

Kurita, K., H. Ichikawa, S. Ishizeki, H. Fujisaki and Y. Iwakura. 1982. "Studies on Chitin. 8. Modification Reaction of Chitin in Highly Swollen State with Aromatic Cyclic Carboxylic Acid Anhydrides," *Makromol. Chem.,* 183:1161.

Kurita, K., S. Inoue and S. Nishimura. 1991. "Preparation of Soluble Chitin Derivatives as Reactive Precursors for Controlled Modifications: Tosyl- and Iodo-chitins," *J. Polym. Sci., Part A: Polym. Chem.,* 29:937.

Kurita, K., H. Yoshino, K. Yokota, M. Ando, S. Inoue, S. Ishii and S. Nishimura. 1992. "Preparation of Tosylchitins as Precursors for Facile Chemical Modifications of Chitin," *Macromolecules,* 25:3786.

Kurita, K., K. Tomita, T. Tada, S. Ishii, S. Nishimura and K. Shimoda. 1993a. "Squid Chitin as a Potential Alternative Chitin Source: Deacetylation Behavior and Characteristic Properties," *J. Polym. Sci., Part A: Polym. Chem.,* 31:485.

Kurita, K., K. Tomita, T. Tada, S. Nishimura and S. Ishii. 1993b. "Reactivity Characteristics of a New Form of Chitosan. Facile *N*-Phthaloylation of Chitosan Prepared from Squid β-Chitin for Effective Solubilization," *Polym. Bull.,* 30:429.

Kurita, K., S. Ishii, K. Tomita, S. Nishimura and K. Shimoda. 1994. "Reactivity Characteristics of Squid β-Chitin as Compared with Those of Shrimp Chitin: High Potentials of Squid Chitin as a Starting Material for Facile Chemical Modifications," *J. Polym. Sci., Part A: Polym. Chem.,* 32:1027.

Nishi, N., J. Noguchi, S. Tokura and H. Shiota. 1979. "Studies on Chitin. I. Acetylation of Chitin," *Polym. J.*, 11:27.

Nishimura, S., H. Kuzuhara, Y. Takiguchi and K. Shimahara. 1989. "Peracetylated Chitobiose: Preparation by Specific Degradations of Chitin and Chemical Manipulations," *Carbohydr. Res.*, 194:223.

Nishimura, S., O. Kohgo, K. Kurita, C. Vittavatvong and H. Kuzuhara. 1990a. "Syntheses of Novel Chitosan Derivatives Soluble in Organic Solvents by Regioselective Chemical Modifications," *Chem. Lett.*, 243.

Nishimura, S., K. Kurita and H. Kuzuhara. 1990b. "A Rapid and Efficient Synthesis of a Trisaccharide Sequence Related to the Core Structure of the Asparagine-Linked Type Glycoproteins by Using a Chitobiose Derivative as a Key Starting Material," *Chem. Lett.*, 1611.

Nishimura, S., K. Matsuoka and K. Kurita. 1990c. "Synthetic Glycoconjugates: Simple and Potential Glycoprotein Models Containing Pendant *N*-Acetyl-D-glucosamine and *N,N'*-Diacetylchitobiose," *Macromolecules*, 23:4182.

Nishimura, S., K. Matsuoka, T. Furuike, S. Ishii, K. Kurita and K. M. Nishimura. 1991a. "Synthetic Glycoconjugates. 2. *n*-Pentenyl Glycosides as Convenient Mediators for the Syntheses of New Types of Glycoprotein Models," *Macromolecules*, 24:4236.

Nishimura, S., O. Kohgo and K. Kurita. 1991b. "Chemospecific Manipulations of a Rigid Polysaccharide: Syntheses of Novel Chitosan Derivatives with Excellent Solubility in Common Organic Solvents by Regioselective Chemical Modifications," *Macromolecules*, 24:4745.

Nishimura, S., Y. Miura, L. Ren, M. Sato, A. Yamagishi, N. Nishi, S. Tokura, K. Kurita and S. Ishii. 1993. "An Efficient Method for the Syntheses of Novel Amphiphilic Polysaccharides by Regio- and Thermoselective Modifications of Chitin," *Chem. Lett.*, 1623.

Osawa, T. 1966. "Lysozyme Substrates. Synthesis of *p*-Nitrophenyl 2-Acetamido-4-*O*-(2-acetamido-2-deoxy-*β*-D-glucopyranosyl)-2-deoxy-*β*-D-glycopyranoside and Its *β*-D-(1→6) Isomer," *Carbohydr. Res.*, 1:435.

Schorigin, P. and E. Hait. 1935. "Acetylation of Chitin," *Chem. Ber.*, 68B:971.

CHAPTER 5

Characterization and Solution Properties of Chitosan and Chitosan Derivatives

M. RINAUDO, M. MILAS, AND J. DESBRIÈRES

CERMAV-CNRS[1]

BP 53

38041 Grenoble Cedex 9, France

INTRODUCTION

Chitin is one of the most abundant natural polymers; many organisms produce this polysaccharide, but the main industrial sources are crab, lobster, or shrimp shells. Chitin represents a family of partially N-acetylated $\beta 1 \rightarrow 4$ D-glucosamine polymers; chitin corresponds to an insoluble material in aqueous solution when the degree of acetylation (DA) is larger than 40–50%. Chitosan is the fully deacetylated chitin, i.e., a pure D-glucosamine polymer, but also polymers with low DA such as to become soluble in acidic conditions. For many years our work has dealt with the chemical modification of chitosan to extend its domains of application; we are also especially involved in the characterization of these chitosans and chitosan derivatives as well as in the study of their solution properties.

EXPERIMENTAL

Different commercial samples were used mainly from Protan; their initial degree of acetylation (DA) was between 10% and 20%. The

---

[1]Affiliated with University Joseph Fourier, Grenoble.

samples are purified easily. First, they are solubilized in acetic acid, filtered, and then the solution is neutralized with NaOH and the chitosan precipitates. It is washed with water and ethanol and then dried under vacuum at ambient temperature. The DA values were determined using ^{1}H NMR, considered to be the most sensitive method [1]; the chitosan is dissolved in D_2O in the presence of HCl and exchanged with D_2O. The DA was determined from the integral of the $-CH_3$ signal at 1.98 ppm compared with the integral of the H-1 proton signals considered as an internal standard. The acetylation degree can also be determined from conductimetric titration: the chitosan sample is dissolved in water with a known quantity of HCl and titrated using a sodium hydroxide solution. The viscometric average molecular weights $\overline{Mv}$ were estimated from the intrinsic viscosity using the relation proposed earlier for DA < 10%:

$$[\eta] = 0.076 \times M^{0.77} \tag{1}$$

after solubilization in the solvent AcOH 0.3 M/AcONa 0.2 M [2]. The viscosity measurements were performed with an Ubbelohde capillary ($\phi = 0.58$ mm) at $25 \pm 0.1°C$. A low shear viscometer (LS 30 from Contraves) was also used to test the effect of the shear rate on viscosity measurements. The steric exclusion chromatography (SEC) experiments were performed using a multidetector system previously described [3]; the solvent for chitosan was AcOH 0.3 M/AcONa 0.2 M. The dn/dc was determined as 0.163 (mL/g).

The carboxylic content in O- and N-carboxymethylchitosan may be determined from conductimetric or turbidimetric titrations using a low molecular weight strong polycation. The carboxymethylchitosan is dissolved in water and titrated with a quaternary ammonium polymer. The substitution degree can also be measured from ^{13}C NMR spectra [1].

RESULTS AND DISCUSSION

Chitosan

Chitin is usually deacetylated in the presence of an excess of sodium hydroxide. To avoid depolymerization, it is important to reduce the reducing end group of the molecule in the presence of $NaBH_4$ and decrease the concentration of NaOH used [4]. We demonstrated that a concentration of NaOH equal to 5% w/v at a temperature of 100°C for 3 h was convenient to get a nearly completely deacetylated sample. Chitosan is soluble in acid medium (pH < 6); the $-NH_2$ groups of the

D-glucosamine unit are important to get solubilization but also for chelating properties [5–8]. The most effective techniques to characterize chitosans are ¹H NMR to determine the residual degree of acetylation (DA) (Figure 1) and SEC for macromolecular characteristics. The DA may also be determined using conductimetric titration. The curve obtained by conductimetry when chitosan dissolved in a known acid content is titrated by sodium hydroxide is given in Figure 2. The DA obtained on an industrial sample from Protan is 12% by ¹H NMR and 12.4% from conductimetric titration.

The molecular weight distribution of the different samples prepared were determined by SEC with multidetection. The most important thing is to get a clean solution of the polymer; the best solvent we proposed was the following: AcOH 0.3 M added in AcONa 0.2 M. The pH of the solution is approximately 4.5. In these conditions the SEC experiments give the average values $\overline{Mw}$, $\overline{Mn}$, $[\eta]$, but also the parameters relating the radius of gyration $\langle \varrho^2 \rangle^{1/2}$ and the intrinsic viscosity $[\eta]$ with the molecular weight M:

$$\langle \varrho^2 \rangle^{1/2} \, (\mathring{A}) = K' M^v \tag{2}$$

$$[\eta] \, (\mathrm{ml/g}) = K M^a \tag{3}$$

From our data, we determined that K' and v are equal to 0.560 and 0.54, respectively; K and a are 0.076 and 0.77.

From these results and following the method discussed in a previous paper [9], we determined the intrinsic persistence length of this molecule, $L_p = 50 \, \mathring{A}$, admitting a worm-like chain behavior [2].

The persistence length L_p was found independent of DA when DA $\leq 20\%$. This value is lower than some data given in the literature [10,11], but these discrepancies are related to the presence of aggregates influencing the light-scattering measurements.

When the L_p values are known (or predicted from a convenient modeling [12,13]), the dimensions of the molecules $\langle \varrho \rangle^{1/2}$ and $[\eta]$ can be predicted with good accuracy when the molecular weight is known. Then, also the specific viscosity of a solution at a given polymer concentration can be determined if we admit for moderate concentration the relation:

$$(\eta - \eta_0)/\eta_0 = C[\eta] + k'(C[\eta])^2 + B(C[\eta])^n \tag{4}$$

with $k' = 0.30$, $B = 0.065$, and $n = 4$.

In Equation (4) η is the viscosity of a solution containing a polymer concentration C (in g/ml), and η_0 is the viscosity of the solvent; η must be taken on the Newtonian plateau.

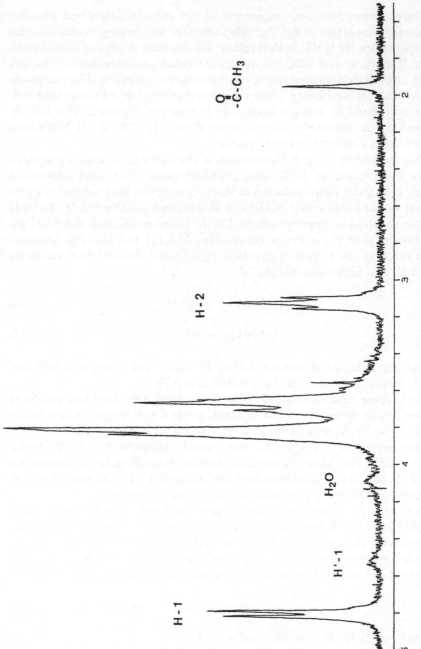

Figure 1. ¹H NMR spectrum of chitosan dissolved in D_2O in the presence of HCl (polymer concentration 10 g/l).

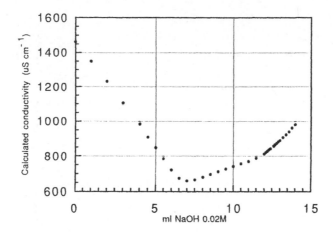

Figure 2. Conductimetric titration of chitosan (determination of DA).

O,N-Carboxymethylchitosan

One of the most important chitin derivatives results from the carboxymethylation of chitosan. This reaction allows to get a water-soluble polymer in a large range of pH and especially over pH 6 where chitosan is no longer soluble.

Two very different reactions were proposed, depending on the position of substitution: *N*-carboxymethylation is easily realized in the presence of glyoxylic acid and the *N*-dicarboxymethylated derivative with a substitution of 95% was described [14]. The structure of the monomeric unit is:

$$R\text{--}N\text{--}CH_2COOH$$
$$|$$
$$CH_2COOH$$

This derivative is perfectly water soluble and can be characterized by ^{1}H and ^{13}C NMR [1] (Figures 3 and 4) and also by SEC.

Considering the multidetection SEC, using a 0.1 M ammonium nitrate solution as eluent, the same treatment as the one performed on chitosan determines the relations $[\eta]$ (M) and $\langle\varrho^2\rangle^{1/2}$ (M):

$$\langle\varrho^2\rangle^{1/2} \text{ (Å)} = 0.235M^{0.57} \tag{5}$$

$$[\eta] \text{ (ml/g)} = 0.051M^{0.72} \tag{6}$$

The exponents are very close to the ones found with chitosan, and

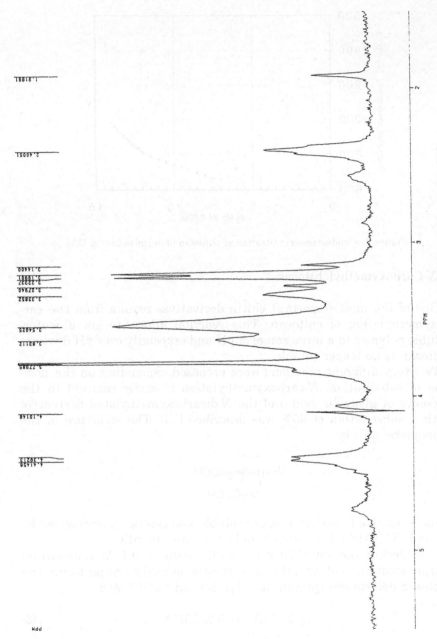

Figure 3. ¹H NMR spectrum of carboxymethylchitosan dissolved in D₂O.

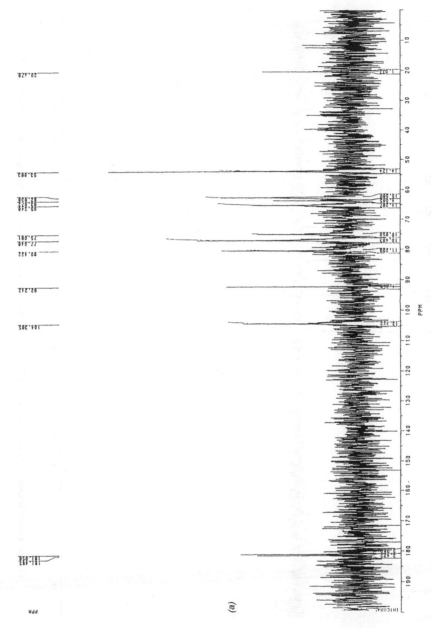

Figure 4. ^{13}C (a) and ^{13}C (DEPT) (b) NMR spectra of carboxymethylchitosan dissolved in D_2O.

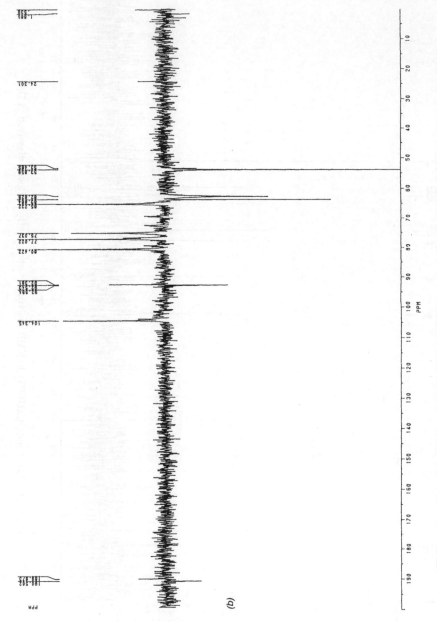

Figure 4 (continued). ^{13}C (a) and ^{13}C (DEPT) (b) NMR spectra of carboxymethylchitosan dissolved in D_2O.

specifically ν is near 0.5. From these data a persistence length of 50 Å can be determined. It is the same as the original chitosan, meaning that the rigidity of the macromolecular chain is the same.

In these conditions, one of the specific properties of chitosan, i.e., chelating properties, progressively disappears when the substitution degree increases. For this reason, it is useful to perform the O-carboxymethylation as discussed previously [14]. We recently reconsidered this modification using the usual reactants NaOH and sodium monochloracetate. The degree of substitution is controlled by the ratio of reactants, time, and temperature of reaction. At low temperature and using these reagents the substitution occurs preferentially on the C-6 and C-3 positions preserving the $-NH_2$ position [15]. The advantage of this method that proceeds in high NaOH concentration is to deacetylate the initial chitosan in the same time giving a more regular polymer derivative. The analysis of the structure is more difficult here because we have $-NH_2$ and carboxylic group to titrate separately in a polyelectrolyte-type system. The specific titration of carboxylic groups is performed by polycationic titration. The principle of this method is the ability of polyanions and polycations to interact by electrostatic forces; the complex formed is generally insoluble at stoichiometry and redissolves with an excess of one of the polyelectrolytes [16,17]. When the temperature of the reaction increases, the substitution occurs on the $-OH$ position as well as on the $-NH_2$ sites. The ^{13}C NMR spectrum of a carboxymethylchitosan is given in Figure 4.

The assignment of different substituent sites is performed using ^{13}C NMR DEPT spectrum and ^{1}H-^{13}C correlations [1]. For quantitative analysis of these spectra, the proton and carbon NMR spin-lattice relaxation times were determined by means of the inversion recovery method (Table 1). Then the experimental conditions used to determine the degree of carboxymethylation and the partial degree of substitution on the different positions (N, C-3, or C-6) need to adopt a delay time large enough; a value of 1.85 second was chosen.

The minimum degree of carboxymethylation to get a good solubility for a pH value larger than 7 must be over 0.5 and preferably larger than 1. Nevertheless, this amphoteric polyelectrolyte presents a lack of solubility in the intermediate range (from 4 to 7). From SEC results a slight chain degradation may be observed during O-carboxymethylation reactions.

N-Alkylation

For grafting hydrophobic alkyl chains, reductive alkylation is chosen as an easy and versatile method to form a covalent bond between an

Table 1. Protons and carbon NMR spin-lattice
relaxation times for carboxymethylchitosan.

	Chemical Displacement (ppm)	T1 (sec)
H-1	4.86	0.64
H-1	4.77	0.70
H-1	4.51	0.74
H-3, H-4, H-5, H-6	3.77	0.65
H-3, H-4, H-5, H-6	3.75	0.60
H-3, H-4, H-5, H-6	3.64	0.63
H-2	3.10	1.10
Acetyl	1.96	1.01

	Chemical Displacement (ppm)	T1 (sec)
C-1	99.2	0.33
C-4	79.2	0.26
C-5	76.85	0.30
C-3	72.2	0.65
Substituted C-6	71.1	0.15
Substituted C-3	69.1	0.41
C-6	62.45	0.17
C-2	58.0	0.34
Substituted N	50.7	0.42
Acetyl	24.3	0.50

alkyl substituent and the amine function of chitosan. The preparation
of these compounds is obtained from the procedure described by
Yalpani and Hall [18] using the aldehyde and sodium cyanoborohydride
at pH 5.5 [19]. The mechanism of this reaction is [20]:

$$\backslash C{=}O + HNR_2 \underset{slow}{\rightleftharpoons} \backslash C\overset{+}{=}\underset{R}{\overset{R}{N}} \xrightarrow[fast]{BH_3CN^-} \backslash HCNR_2$$

The degree of alkylation is determined by ¹H NMR in acidic conditions when the polymer is soluble (Figure 5). The presence on the
macromolecular chain of hydrophobic substituents gives an amphiphilic derivative soluble in acidic medium. The alkyl substituents
cause hydrophobic interaction and as first characteristics an increase of
the viscosity of aqueous solutions (Figure 6) [19]. In that respect the
balance between electrostatic repulsion of the cationic polymer (related

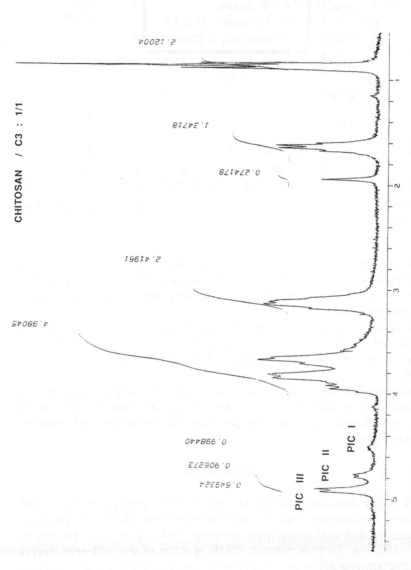

Figure 5. ^{1}H NMR spectrum of alkylated chitosan dissolved in D_2O in the presence of HCl (monomolar ratio of chitosan/C_3 aldehyde being 1/1).

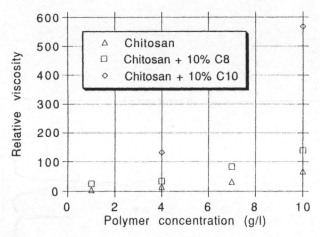

Figure 6. Variation of relative viscosity with polymer concentration for alkylated chitosans (solvent AcOH 0.2 M).

to pH) and the hydrophobic interaction is important to control the solution properties. The degree of substitution and the length of the alkyl chain can easily be adjusted, the solubility being very sensitive to these parameters.

The solubility is directly related to the number and the length of the substituted alkyl chain. For example, with a C_8 alkyl chain the chitosan derivative is soluble in AcOH 0.2 M up to a weight concentration of 20%.

Quaternized chitosan derivatives were also previously prepared and characterized as described in Reference [14]. The experimental conditions are chosen for reducing polymer degradation and avoiding O-methylation.

CONCLUSION

This paper describes mainly the methods to prepare water-soluble chitosan derivatives and to characterize the different products. Specifically, different techniques are described to determine the degree of substitution. In that respect, NMR spectra of the different derivatives are analyzed.

The macromolecular characteristics are completely determined using steric exclusion chromatography with a multidetection; the role of solvent is of major importance in this determination. The main properties of these chitosan and chitosan derivatives are their thickening properties and also their flocculating [21–22] and chelating properties.

REFERENCES

1. Rinaudo, M., Le Dung, P., Gey, C. and Milas, M. 1992. "Substituent Distribution on *O,N*-Carboxymethylchitosans by ¹H and ¹³C n.m.r.," *Int. J. Biol. Macromol.*, 14:122–128.

2. Rinaudo, M., Milas, M. and Le Dung, P. 1993. "Characterization of Chitosan. Influence of Ionic Strength and Degree of Acetylation on Chain Expansion," *Int. J. Biol. Macromol.*, 15:281–285.

3. Tinland, B., Rinaudo, M. and Mazet, J. 1988. "Characterization of Water-Soluble Polymers by Multidetection Size-Exclusion Chromatography," *Makromol. Chem., Rapid Commun.*, 9:69–73.

4. Rinaudo, M., Le Dung, P. and Milas, M. February 1993. French Patent FR 9302182.

5. Muzzarelli, R. A. A. 1973. *Natural Chelating Polymers*, Pergamon, Oxford.

6. Masri, M. S., Reuter, F. W. and Friedman, M. 1974. "Binding of Metal Cations by Natural Substances," *J. Appl. Polym. Sci.*, 18:675–681.

7. Eiden, C. A., Jewell, C. A. and Wightman, J. P. 1980. "Interaction of Lead and Chromium with Chitin and Chitosan," *J. Appl. Polym. Sci.*, 25:1587–1599.

8. Hall, L. D. and Yalpani, M. 1980. "Enhancement of the Metal-Chelating Properties of Chitin and Chitosan," *Carbohydr. Res.*, 83:C5–C7.

9. Chazeau, L., Milas, M. and Rinaudo, M. 1995. "Conformations of Xanthan in Solution Analysis by Steric Exclusion Chromatography," *Int. J. of Polym. Anal. Characterization*, 2:27–29.

10. Rinaudo, M. and Domard, A. 1988. "Solution Properties of Chitosan," *4th International Conference on Chitin and Chitosan* (1988, Trondheim), *Chitin and Chitosan*, Ed. Skjak-Braek, Anthonsen, Sandford, Elsevier, pp. 71–86.

11. Terbojevich, M., Cosani, A., Conio, G., Marsano, E. and Branchi E. 1991. "Chitosan: Chain Rigidity and Mesophase Formation," *Carbohydr. Res.*, 209:251–260.

12. Mazeau, K., Tvaroska, I., Rinaudo, M. and Milas, M. 1995. unpublished results.

13. Yui, T., Kobayashi, H., Kitamura, S. and Imada, K. 1994. "Conformational Analysis of Chitobiose and Chitosan," *Biopolymers,* 34:203–208.

14. Le Dung, P., Milas, M., Rinaudo, M. and Desbrières, J. 1994. "Water Soluble Derivatives Obtained by Controlled Chemical Modifications of Chitosan," *Carbohydr. Polym.*, 24:209–214.

15. Rinaudo, M., Le Dung, P. and Milas, M. 1992. "A New and Simple Method of Synthesis of Carboxymethylchitosan," *5th International Conference of Chitin and Chitosan* (1991, Princeton), *Advances in Chitin and Chitosan,* Brine, Sandford and Zitakis, eds., Elsevier, pp. 516–525.

16. Desbrières J. 1980. "Polyanion-Polycation Interactions. Applications to Ultrafiltration," Thesis, Grenoble.

17. Desbrières, J. and Rinaudo, M. 1981. "Formation of Polyelectrolyte Complexes in an Organic Solvent," *Eur. Polym. J.,* 17:1265–1269.

18. Yalpani, M. and Hall, L. D. 1984. "Some Chemical and Analytical Aspects of Polysaccharide Modifications. 3. Formation of Branched-Chain, Soluble Chitosan Derivatives," *Macromolecules,* 17:272–281.

19. Desbrières, J., Martinez, C. and Rinaudo, M. 1996. "Hydrophobic Derivatives of Chitosan: Characterization and Rheological Behaviour," *Int. J. Biol. Macromol.,* 19:27–28.

20. Borch, R. F., Bernstein, M. D. and Durst, H. D., 1971. "The Cyanohydridoborate Anion as a Selective Reducing Agent," *J. Am. Chem. Soc.,* 93:2897–2904.

21. Terrassin, C., 1986. "Synthesis and Characterization of Water Soluble Polymers. Interactions with Elements of a Fibrous Paper Manufacturer Suspension," Thesis, Grenoble.

22. Domard, A., Rinaudo, M. and Terrassin, C. 1989. "Adsorption of Chitosan and a Quaternized Derivative on Kaolin," *J. Appl. Polym. Sci.,* 38:1799–1806.

CHAPTER 6

Soluble Precursors for Efficient Chemical Modifications of Chitin and Chitosan

KEISUKE KURITA

Department of Industrial Chemistry
Faculty of Engineering
Seikei University
Musashino-shi, Tokyo 180, Japan

INTRODUCTION

Chitin is characterized by specific properties including bioactivity, biocompatibility, and biodegradability. Moreover, it is potentially more useful than cellulose for developing advanced functions because of the presence of amino groups. Chitin is thus attracting a great deal of attention not only as an unutilized biomass resource but as a novel type of specialty functional material.

Chitin is reported to be soluble in N,N-dimethylacetamide (DMAc) containing lithium chloride, but the extent of solubility is dependent on the origins of chitin (Rutherford and Austin, 1978). Although special attention has been paid to chemical modifications of chitin to enable full exploration of the high potentials, modification reactions of chitin are generally difficult owing to the lack of solubility. Reactions under heterogeneous conditions are accompanied by various problems, including low extents of reaction, difficulty in regioselective substitution, structural nonuniformity of the products, and partial degradation due to harsh reaction conditions. Many efforts have thus been focused on the development of facile modification reactions to prepare derivatives with well-defined structures. This would open the way to extensive utilizations in the near future.

103

Of various attempts to develop solubility of this rigid intractable polysaccharide, destruction of the crystalline structure by incorporating appropriate substituents has been effective. This chapter deals with the preparation of soluble derivatives such as water-soluble chitin, tosyl-chitin, iodo-chitin, and N-phthaloyl-chitosan, and some efficient chemical modifications of these precursors in solution.

WATER-SOLUBLE CHITIN

Preparation

Chitin can be dissolved in aqueous alkali by steeping in concentrated sodium hydroxide followed by treating with crushed ice to give an alkali chitin solution. Deacetylation proceeds efficiently under these homogeneous conditions, and for instance, an alkali chitin solution (1% chitin, 10% sodium hydroxide) left for 70 h at room temperature gives a product with about 50% deacetylation (Sannan et al., 1975, 1976):

Water-soluble chitin (l ≈ m)

(1)

The resulting product is soluble in neutral water. Higher or lower deacetylations fail to lead to complete solubilization, and the degree of deacetylation should be 0.45–0.55. The water solubility is ascribable to the greatly enhanced hydrophilicity as a result of random distribution of acetyl groups at about half of the amino groups (Kurita et al., 1977b). The products with 50% deacetylation prepared by the conventional hydrolysis under heterogeneous conditions are not soluble.

Alternatively, random N-acetylation of chitosan to a degree of acetylation of approximately 0.5 also gives rise to the water-soluble chitin [Equation (1)] (Kurita et al., 1989, 1991a). Although this method uses chitosan as a starting material, it may be superior to the above-mentioned deacetylation of chitin in that the isolation of the product is easier, and moreover, the water-soluble chitin with a desired molecular weight can be prepared starting from an appropriate chitosan.

Chemical Modifications

The water-soluble chitin is useful as a reactive precursor for controlled modifications, since the reactions can be performed in a homogeneous aqueous solution or in a highly swollen state in organic solvents. Some modification reactions that have been difficult owing to the insoluble nature of chitin have thus become possible.

Efficient acylation is, for example, possible in aqueous media or in organic solvents. When the water-soluble chitin is treated with acetic acid and dicyclohexylcarbodiimide in a mixed solvent of water and *N,N*-dimethylformamide (DMF), acetylation takes place selectively at the free amino groups to give fully *N*-acetylated pure chitin, poly(*N*-acetyl-D-glucosamine) (Kurita et al., 1977a).

Acylation reactions of the water-soluble chitin can also be carried out in a highly swollen state in organic solvents when aqueous media are not suitable. A highly swollen gel, obtained by pouring an aqueous solution of the water-soluble chitin into organic solvents, such as pyridine, DMAc, and dimethyl sulfoxide (DMSO), allows many kinds of modification reactions to proceed under almost homogeneous conditions. Acetylation of the free amino groups with acetic anhydride is particularly facile, and even hydroxyl groups are acetylated to some extent in a few minutes at room temperature (Kurita et al., 1977a).

Aromatic cyclic acid anhydrides, such as phthalic anhydride, trimellitic anhydride, and pyromellitic dianhydride, give amic acid derivatives smoothly. On heating, the amic acids are converted into the corresponding imide derivatives by dehydration cyclization:

$$ Water\text{-}soluble\ chitin\ (l \approx m) $$

(2)

The imide derivatives prepared from the latter two anhydrides show reactivity toward nucleophiles owing to the presence of the carboxyl and acid anhydride groups. It is noteworthy that the derivative from phthalic anhydride exhibits solubility in DMSO (Kurita et al., 1982).

Nicotinoylation with nicotinic anhydride is also achieved with the water-soluble chitin in a swollen state in pyridine to give N-nicotinoyl-chitin, which is then quaternized and reduced to incorporate dihydro-nicotinamide groups, which are the active site of coenzyme NADH:

(3)

The resulting derivatives are polymeric asymmetric reducing agents and afford ethyl mandelate from ethyl benzoylformate with ee 33% (Kurita et al., 1986a).

Among some chemical modifications that the water-soluble chitin has made possible, acylation with α-amino acid N-carboxy anhydrides (NCAs) is one of the most important reactions, and this has enabled the introduction of polypeptide side chains (see Chapter 17 in this book). Aldehydes give Schiff base derivatives of the water-soluble chitin (Kurita et al., 1988). Low extents of cross-linking with glutaraldehyde improve the adsorption capacity for metal cations (Kurita et al., 1986b).

TOSYL- AND IODO-CHITINS

Preparation

Chitin and chitosan are much less accessible to potential reactants than cellulose, and it is difficult to prepare derivatives with well-defined structures owing to their strong intermolecular forces and highly crystalline structures. Development of organosoluble precursors are thus highly desirable to enable controlled modification reactions in solution under mild conditions.

For improving reactivity as well as solubility in organic solvents, introduction of *p*-toluenesulfonyl (tosyl) or iodo groups into chitin is considered to be promising, since these groups are expected to be bulky enough for developing solubility, and furthermore, the derivatives would be highly reactive.

Tosylation of the ordinary α-chitin is quite sluggish in pyridine, even at elevated temperatures, probably because of the heterogeneous reaction conditions. It is, however, accomplished efficiently by the interfacial reaction between an aqueous alkali chitin solution and a chloroform solution of tosyl chloride at 0°C (Kurita et al., 1991b, 1992):

$$(4)$$

The tosylation degree increases with an increase in the amount of tosyl chloride, but levels off at approximately 1.0, indicating the preferential substitution at the less hindered C-6 positions as in the case of carboxymethylation. Chitin is deacetylated to some extent in nature, and moreover, partial deacetylation takes place during the tosylation reaction because of the alkaline conditions. To prepare tosyl-chitins with well-defined structures, therefore, subsequent *N*-acetylation is necessary, but furthermore, a small amount of *O*-acetyl groups formed at the same time have to be removed. β-Chitin, however, shows higher reactivity than the ordinary α-chitin and can be tosylated in pyridine at room temperature (see Chapter 4 in this book).

The resulting tosyl-chitins are easily converted into the corresponding iodo-chitins when treated with sodium iodide in DMSO solution at 85°C [Equation (4)].

The tosyl-chitins are almost soluble in water when the substitution degrees are low but become less hydrophilic when the substitution degree increases. The derivatives with a substitution degree above 0.5 are soluble in organic solvents such as DMSO and DMAc. The derived iodo-chitins show even better solubility in organic solvents.

Chemical Modifications

Owing to the high solubility and reactivity of tosyl- and iodo-chitins, they are promising as reactive precursors for various modification reactions in homogeneous solution in organic solvents.

Reduction of tosyl-chitins with sodium borohydride, for example, gives 6-deoxy derivatives in DMSO [Equation (4)] (Kurita et al., 1992). They are also converted into mercapto-chitins as shown in Equation (4). Tosyl-chitins are first treated with potassium thioacetate in DMSO solution, and subsequent S-deacetylation with methoxide results in the formation of mercapto-chitins (Kurita et al., 1993). The products are interesting because of the presence of reactive and biologically significant mercapto groups.

Biodegradability is one of the most important aspects of chitin derivatives, and the mercapto-chitins have turned out to be more susceptible to lysozyme than the original chitin. This is interpreted in terms of the loose arrangement of the chitin molecules as a result of the introduction of bulky mercapto groups.

Iodo-chitins are also reactive precursors, and efficient graft copolymerization of styrene is possible by both cationic and radical mechanisms (see Chapter 17 in this book).

N-PHTHALOYL-CHITOSAN

Preparation

As suggested by the solubility of N-phthaloylated water-soluble chitin in DMSO (Kurita et al., 1982), the phthaloyl group is effective to impart solubility owing to the bulky nature and complete removal of N-attached hydrogens that form intermolecular hydrogen bonds. This has led to the preparation of N-phthaloylated chitosan, which shows even better solubility in organic solvents.

The reaction is accomplished by heating suspended chitosan with excess phthalic anhydride in DMF at 130°C (Nishimura et al., 1990):

$$(5)$$

As N-phthaloylation proceeds, the mixture becomes a homogeneous solution. It is noteworthy that the reaction proceeds much more facilely with chitosan derived from squid β-chitin than the ordinary one from shrimp α-chitin. The resulting fully N-phthaloylated chitosan is readily soluble even in pyridine in addition to polar solvents such as DMF, DMAc, and DMSO.

Chemical Modifications

High solubility of N-phthaloyl-chitosan has enabled various modification reactions to proceed quantitatively in homogeneous solution in organic solvents suitable for reactions. The preparative procedures based on N-phthaloyl-chitosan are useful for regioselective and quantitative introduction of substituents. After desired modification reactions, the phthaloyl groups are easily removed with hydrazine to regenerate the free amino groups [Equation (5)] (Nishimura et al., 1990, 1991).

Tritylation, for example, takes place at the C-6 hydroxyl groups quantitatively in pyridine solution to give the 2-N-phthaloyl-6-O-trityl derivative. Subsequent dephthaloylation gives 6-O-trityl-chitosan, which is a suitable precursor for introducing substituents at the C-2 amino and C-3 hydroxyl groups. N-Acetylation of 6-O-trityl-chitosan with acetic anhydride in methanol gives 6-O-trityl-chitin having free hydroxyl groups at C-3.

Acylation of 6-O-trityl-chitosan with excess palmitoyl chloride in pyridine at 45°C gives the N,O-dipalmitoyl derivative. At 70°C, however, triacylation takes place to give the N,N,O-tripalmitoyl derivative. Detritylation of the resulting products with dichloroacetic acid regenerates the free hydroxyl groups at C-6. All these reactions proceed quantitatively in solution. The free hydroxyl groups are then sulfated with sulfur trioxide/trimethylamine in pyridine to introduce hydrophilic sulfate groups, leading to amphiphilic chitin derivatives (Nishimura et al., 1993):

(6)

The tripalmitoylated derivative shows particularly high solubility even in low-boiling solvents and is capable of forming liposomes and Langmuir monolayers.

Acetylation of the C-3 hydroxyl groups of 2-*N*-phthaloyl-6-*O*-trityl-chitosan in pyridine in the presence of 4-dimethylaminopyridine gives rise to fully substituted chitosan. Detritylation affords the derivative having free hydroxyl groups at C-6, the other groups being protected. The C-6 hydroxyl groups can be trimethylsilylated to further improve solubility. The resulting 3-*O*-acetyl-6-*O*-trimethylsilyl-2-*N*-phthaloyl-chitosan is soluble even in dichloromethane and is useful for introducing substituents regioselectively at C-6 (Kurita et al., 1994).

When it is subjected to the glycosidation reaction of a mannose orthoester in the presence of trimethylsilyl trifluoromethanesulfonate in dichloromethane, α-mannoside groups are introduced at C-6 to give nonnatural branched polysaccharides. Subsequent deprotection results in the formation of branched chitosans. Finally, selective *N*-acetylation gives chitins having α-mannoside groups at C-6:

$$
\left[\begin{array}{c}\text{OTr}\\ \text{AcO}\overset{\displaystyle O}{\underset{\text{NPhth}}{\rule{0pt}{0pt}}}\text{O}-\end{array}\right]_n \xrightarrow{\text{DCA}} \left[\begin{array}{c}\text{OH}\\ \text{AcO}\overset{\displaystyle O}{\underset{\text{NPhth}}{\rule{0pt}{0pt}}}\text{O}-\end{array}\right]_n \xrightarrow[\text{Me}_3\text{SiCl}]{(\text{Me}_3\text{Si})_2\text{NH}} \left[\begin{array}{c}\text{OSiMe}_3\\ \text{AcO}\overset{\displaystyle O}{\underset{\text{NPhth}}{\rule{0pt}{0pt}}}\text{O}-\end{array}\right]_n \xrightarrow{\text{TMSOTf}}
$$

$$
\left[\begin{array}{c}\text{OAc}\\ \text{OAc}\\ \text{AcO}\\ \text{AcO}\\ \text{AcO}\overset{\displaystyle O}{\underset{\text{NPhth}}{\rule{0pt}{0pt}}}\text{O}-\end{array}\right]_n \xrightarrow[\text{2) N}_2\text{H}_4\cdot\text{H}_2\text{O}]{\text{1) 1N NaOH}} \left[\begin{array}{c}\text{OH}\\ \text{OH}\\ \text{HO}\\ \text{HO}\\ \text{HO}\overset{\displaystyle O}{\underset{\text{NH}_2}{\rule{0pt}{0pt}}}\text{O}-\end{array}\right]_n \xrightarrow{\text{Ac}_2\text{O}} \left[\begin{array}{c}\text{OH}\\ \text{OH}\\ \text{HO}\\ \text{HO}\\ \text{HO}\overset{\displaystyle O}{\underset{\text{NHAc}}{\rule{0pt}{0pt}}}\text{O}-\end{array}\right]_n \quad (7)
$$

The resulting branched chitosans and chitins are readily soluble in water and show specific affinity for concanavalin A owing to the presence of the α-mannoside branches (Kurita et al., 1994).

Several other modification reactions based on *N*-phthaloyl-chitosan have also been confirmed to be quite facile and quantitative, unlike the conventional sluggish reactions on chitin and the derivatives.

CONCLUSION

Although various possibilities of utilization of chitin have been suggested, there still remain many problems to solve. The biopolymer's intractable nature is undoubtedly responsible for the tardy progress in application studies. Much attention, however, has been paid to the development of regioselective and quantitative modification reactions that will make it possible to design desired molecular structures. These controlled modifications will become increasingly important to impart advanced functions and thereby to develop applications in various fields.

REFERENCES

Kurita, K., T. Sannan and Y. Iwakura. 1977a. "Studies on Chitin. 3. Preparation of Pure Chitin, Poly(*N*-acetyl-D-glucosamine), from the Water-Soluble Chitin," *Makromol. Chem.,* 178:2595.

Kurita, K., T. Sannan and Y. Iwakura. 1977b. "Studies on Chitin. 4. Evidence for Formation of Block and Random Copolymers of *N*-Acetyl-D-glucosamine and D-Glucosamine by Hetero- and Homogeneous Hydrolyses," *Makromol. Chem.,* 178:3197.

Kurita, K., H. Ichikawa, S. Ishizeki, H. Fujisaki and Y. Iwakura. 1982. "Studies on Chitin. 8. Modification Reaction of Chitin in Highly Swollen State with Aromatic Cyclic Carboxylic Acid Anhydrides," *Makromol. Chem.,* 183:1161.

Kurita, K., Y. Koyama, K. Murakami, S. Yoshida and N. Chau. 1986a. "Studies on Chitin. XII. Chitin Derivatives Having 1,4-Dihydronicotinamide Groups for Asymmetric Reduction," *Polym. J.,* 18:673.

Kurita, K., Y. Koyama and A. Taniguchi. 1986b. "Studies on Chitin. IX. Cross-linking of Water-Soluble Chitin and Evaluation of the Products as Adsorbents for Cupric Ion," *J. Appl. Polym. Sci.,* 31:1169.

Kurita, K., M. Ishiguro and T. Kitajima. 1988. "Studies on Chitin. 17. Introduction of Long Alkylidene Groups and the Influence on the Properties," *Int. J. Biol. Macromol.,* 10:124.

Kurita, K., Y. Koyama, S. Nishimura and M. Kamiya. 1989. "Facile Preparation of Water-Soluble Chitin from Chitosan," *Chem. Lett.,* 1597.

Kurita, K., M. Kamiya and S. Nishimura. 1991a. "Solubilization of a Rigid Polysaccharide: Controlled Partial *N*-Acetylation of Chitosan to Develop Solubility," *Carbohydr. Polym.,* 16:83.

Kurita, K., S. Inoue and S. Nishimura. 1991b. "Preparation of Soluble Chitin Derivatives as Reactive Precursors for Controlled Modifications: Tosyl- and Iodo-Chitins," *J. Polym. Sci., Part A: Polym. Chem.,* 29: 937.

Kurita, K., H. Yoshino, K. Yokota, M. Ando, S. Inoue, S. Ishii and S. Nishimura. 1992. "Preparation of Tosylchitins as Precursors for Facile Chemical Modifications of Chitin," *Macromolecules,* 25:3786.

Kurita, K., H. Yoshino, S. Nishimura and S. Ishii. 1993. "Preparation and Biodegradability of Chitin Derivatives Having Mercapto Groups," *Carbohydr. Polym.,* 20:239.

Kurita, K., M. Kobayashi, T. Munakata, S. Ishii and S. Nishimura. 1994. "Synthesis of Non-Natural Branched Polysaccharides. Regioselective Introduction of α-Mannoside Branches into Chitin," *Chem. Lett.,* 2063.

Nishimura, S., O. Kohgo, K. Kurita, C. Vittavatvong and H. Kuzuhara. 1990. "Syntheses of Novel Chitosan Derivatives Soluble in Organic Solvents by Regioselective Chemical Modifications," *Chem. Lett.,* 243.

Nishimura, S., O. Kohgo and K. Kurita. 1991. "Chemospecific Manipulations of a Rigid Polysaccharide: Syntheses of Novel Chitosan Derivatives with Excellent Solubility in Common Organic Solvents by Regioselective Chemical Modifications," *Macromolecules,* 24:4745.

Nishimura, S., Y. Miura, L. Ren, M. Sato, A. Yamagishi, N. Nishi, S. Tokura, K. Kurita and S. Ishii. 1993. "An Efficient Method for the Syntheses of Novel Amphiphilic Polysaccharides by Regio- and Thermoselective Modifications of Chitin," *Chem. Lett.,* 1623.

Rutherford, P. A. and P. R. Austin. 1978. "Marine Chitin Properties and Solvents," *Proc. Int. Conf. Chitin/Chitosan, 1st,* p. 182.

Sannan, T., K. Kurita and Y. Iwakura. 1975. "Studies on Chitin. 1. Solubility Change by Alkaline Treatment and Film Casting," *Makromol. Chem.,* 176:1191.

Sannan, T., K. Kurita and Y. Iwakura. 1976. "Studies on Chitin. 2. Effect of Deacetylation on Solubility," *Makromol. Chem.,* 177:3589.

FOOD AND AGRICULTURE

CHAPTER 7

Chitosans as Dietary Food Additives

RICCARDO A. A. MUZZARELLI
Faculty of Medicine
University of Ancona
Via Ranieri 67
IT-60100 Ancona, Italy

MASSIMO DE VINCENZI
Department of Metabolism and Pathological Biochemistry
Istituto Superiore di Sanità
IT-00100 Rome, Italy

HYPOCHOLESTEROLEMIC ACTION ON ANIMALS

The nondigestibility in the upper gastrointestinal tract, high viscosity, polymeric nature and high water-binding properties, together with low water binding in the lower gastrointestinal tract, are all responsible for the effective hypocholesterolemic potential of dietary fiber. Chitosan meets most of these criteria and has a highly characteristic property in relation to other dietary plant fibers: it can bind in vitro a variety of anions such as bile acids or free fatty acids at low pH by ionic bonds resulting from its amino group.

Chitosan binds fatty acids to form the corresponding complex salts. The binding is mainly of ionic nature; in fact, the salt is prepared by neutralization of chitosan with edible fatty acids, such as oleic, linoleic, palmitic, stearic, and linolenic acids. The resulting salts, after ingestion by a mammal, bind additional lipids, probably because of hydrophobic interactions (triglycerides, fatty and bile acids, cholesterol, and other sterols), and a great portion of these bound lipids are excreted rather than absorbed by the mammals. Hydrochloric acid

115

in the stomach would not hydrolyze chitosan fatty acid salts because this material would not wet. Rather, this chitinous material would grow in size as it travels through the gastrointestinal tract and would bind additional amounts of lipids. Bound triglycerides would escape hydrolysis by lipase, promoting the excretion of fatty materials, including cholesterol, sterols, and triglycerides (Furda, 1980).

Authors have compared chitosan with cholestyramine, an anionic synthetic resin (Dowex A-1) that is being used to sequester phospholipids, monoglycerides, fatty acids, and cholesterol. This resin might interfere with lipid absorption not only by removal of amphipathic micellar components, such as bile acids and phospholipids, but also by sequestration of cholesterol and products of triglyceride lipolysis.

Cholestyramine is currently used successfully to treat hypercholesterolemia in humans. Although very effective, its use has been questioned because of reports of a possible link to colon cancer in humans and rat experimental models. Cholestyramine sequesters bile salts in rats and alters intestinal morphology in humans. Furthermore, there is a significant increase in rat intestinal tumor induction by various agents when cholestyramine is added to the diet. Diethylaminoethyl (DEAE)-dextran was also effective in reducing intestinal fat absorption and speeding up bile turnover (Pupita and Barone, 1987).

Chitosan, in contrast to cholestyramine, has viscous properties more like those of viscous dietary fiber, such as pectin and guar gum. The viscous fiber also sequesters micellar components in vitro, albeit with considerably less avidity than does cholestyramine and other commercial anion exchangers. This is likely due to a trapping in the gel matrix and may readily dissociate under in vivo conditions (Vahouny et al., 1983).

The available in vivo data indicate that the mechanisms by which cholestyramine and chitosan affect lipid absorption may be different. Sugano et al. (1988) reported that after a 20-day feeding of 5% cholestyramine or chitosan to rats on a cholesterol-containing diet, plasma and liver cholesterol levels in both groups were significantly lower than in controls. However, fecal neutral sterols were elevated only in the chitosan-fed rats, suggesting a difference in the action of chitosan and cholestyramine. This was supported in a subsequent study in which the feeding of chitosan caused increased fecal output of neutral sterols but not acidic steroids, whereas cholestyramine feeding increased mainly fecal acidic steroids. The data of Sugano et al. (1988) also suggest that the mechanism of chitosan action may not be completely parallel with those of the viscous dietary fibers. The water-soluble pectins, when fed at sufficient levels, cause some increase in the output of fecal neutral and acidic steroids, but this appears insufficient to account for their hypocholesterolemic effects. However, chitosan, which has greater hypocholesterolemic potency than pectins in cholesterol-fed rats, does not effectively increase bile acid excretion.

In an animal model, chitosan inhibited the transformation of cholesterol to coprostanol. Increased fiber consumption did not increase fecal excretion of bile acids but caused a marked change in fecal bile acid composition. The pH of the cecum and colon became elevated by chitosan feeding, which affected the conversion of primary bile acids to secondary bile acids in the large intestine. In the cecum, the lithocholic acid concentration increased. These data obtained by Fukada et al. (1991) indicate that chitosan affects the metabolism of intestinal bile acids in rats. Thus, based on evidence derived from analyses of fecal acidic and neutral steroid output, one cannot easily assign a common mechanism of action for cholestyramine, chitosan, and viscous dietary fibers.

Two important parameters used to characterize chitosans are the degree of acetylation and the degree of polymerization. They influence the viscosity of the solutions of chitosan salts and have had prominent roles in the biochemical significance of chitosan. Therefore, the relationship between hypocholesterolemic efficacy and average molecular weight of chitosan was studied in rats fed a cholesterol-enriched (0.5%) diet by Sugano et al. (1988). Several chitosan preparations, with a comparable degree of deacetylation but differing widely in average molecular weight and viscosity, almost completely prevented the rise of serum cholesterol at the 5% dietary level. At the 2% level, chitosans with viscosities at both extremes exerted a comparable cholesterol-lowering action. The results indicate that the hypocholesterolemic action of chitosans is independent of their molecular weight within the tested viscosity range. Judging from the ineffectiveness of the glucosamine, it was suggested that some degree of polymerization is required to provoke a cholesterol-lowering activity (Sugano et al., 1992).

Ikeda et al. (1993) also addressed the subject of the significance of the degree of polymerization of chitosan. The concentration of serum cholesterol at day 7 was significantly lower in rats fed chitosan preparations with MW 5,000–50,000 Da than in those fed cellulose. The 2,000-Da chitosan preparation did not show a hypocholesterolemic effect at day 7. All chitosan hydrolysates, except for 2,000-Da, significantly increased fecal excretion of neutral steroids as cholesterol and coprostanol. The hydrolysates with average molecular weights above 10,000 Da were more effective in enhancing fecal excretion of neutral steroids. The effect on the composition of neutral steroids was diverse depending on the preparations. Excretion of total acidic steroids was slightly increased in rats fed chitosan hydrolysates, except for the 2,000-Da group.

Lymphatic absorption of radioactive cholesterol was demonstrated. The rats given cellulose most effectively absorbed cholesterol during 24 h. Cholesterol absorption in the 2,000-Da group was comparable to the cellulose group at 24 h after administration, but it was lower at 6 and 9 h. The lowering effects of 5,000, 20,000, and 50,000-Da chitosans on cholesterol absorption were comparable and lasted for 24 h.

In particular, Ikeda et al. (1993) noted that the influence of depolymerized chitosan on lipid absorption as a cationic substance is not unequivocal and it is not obvious whether the mechanism proposed for viscous fibers, i.e., the reduced diffusion of micellar lipids to intestinal wall, is applicable to depolymerized chitosans.

Deuchi et al. (1994) investigated the effects of various dietary fibers on the apparent fat digestibility by rats fed on a high-fat diet. Each of twenty-three different fibers was added (5% w/w) to a purified diet containing corn oil (20% w/w). The rats were fed these diets for 2 weeks, and the feces were collected from each animal during the last 3 days. When compared with cellulose (control), ten of the tested fibers significantly increased the fecal lipid excretion. Among these fibers, chitosan markedly increased the fecal lipid excretion and reduced the apparent fat digestibility to about half of the control. The fatty acid composition of the fecal lipids closely reflected that of the dietary fat. These results suggest that chitosan has potency for interfering with fat digestion and absorption in the intestinal tract and for facilitating the excretion of dietary fat into the feces.

Nauss et al. (1983) suggested that a soluble form of chitosan would be able to interfere with intraluminal lipid absorption through the interaction with micelle formation or emulsification of lipids in the enteric phase. Ikeda et al. (1993) offered a similar suggestion on the basis of their study on fully emulsified oil with chitosan and sodium taurocholate. On the other hand, it was also suggested that dietary fiber would not necessarily limit the in vivo absorption. It is therefore noteworthy that chitosan had strong ability to increase the fecal lipid excretion in vivo.

Deuchi et al. (1994) confirmed that cholestyramine did not affect the fecal lipid excretion and that kapok and polypropylene, both capable of forming a hydrophobic environment in the intestinal tract, also had little effect. These results may suggest that chitosan elicited its effect by its unique behavior rather than by any ability to combine with bile acids. With regard to the mechanism, Deuchi et al. (1994) consider that the chitosan dissolved in the stomach forms an emulsion with intragastric oil droplets and would begin to precipitate in the small intestine at pH 6.0–6.5. As the polysaccharide chains start to aggregate, they would entrap fine oil droplets in their matrices, pass through the lumen, and empty into the feces.

Rats were fed by Kanauchi et al. (1994) on high-fat diets containing cellulose (control), chitosan, chitosan ascorbate, chitosan lactate, and chitosan citrate salts. The presence of ascorbate caused a larger increase in the fecal fat excretion than otherwise, without considerably affecting the apparent protein digestibility. The mechanism for depressing fat digestibility by chitosan ascorbate was explained in the following terms: gastric acid-soluble chitosan is mixed with dietary fat in the stomach, the emulsifying process being effectively mediated by ascorbic acid. When the digest emptied from the stomach comes into contact with pancreatic juice (in an alkaline pH range), oil droplets become embedded in the gelled chitosan matrices and are excreted into the feces without fully undergoing absorption. This is supported by the fatty acid composition of the fecal lipid being very similar to that of corn oil in the diet.

Chitosan could also affect the growth rate. This mainly depended on reducing the energy intake than by the inhibition of lipid digestibility. About 80% of dietary fat could not be digested by supplementing chitosan with ascorbic acid, so the rats could not take in sufficient energy when compared with the control animals. Despite feeding with a high-fat diet, the plasma triglyceride concentration and epididymal fat pad weight in the chitosan-receiving groups remained at low levels, probably because of a depressed fat intake (Kanauchi et al., 1994).

If the effect of chitosan on the inhibition of lipid digestion depended only on the foregoing mechanism, it could be seen that chitosan may envelop all nutrients. However, the protein digestibility in rats fed with the chitosan diets was not markedly changed. This recent finding (Deuchi et al., 1994; Kanauchi et al., 1994) could possibly be explained in the light of the effect of chitosan on the permeability of monolayers of intestinal epithelial cells (Artursson et al., 1994). Because of their positive charge, cationic macromolecules, such as protamine, polylysine, and chitosan, can interact with the anionic components (sialic acid) of the glycoproteins on the surface of the epithelial cells. In a similar study an intravenous infusion of protamine produced large increases in lymph flow and in lymph protein clearance. This enhanced transcapillary protein flux was primarily due to the neutralization of fixed anionic sites on the capillary wall. Further, it has been suggested that cationic macromolecules are able to displace cations from electronegative sites (such as tight junctions) on a membrane that require coordination with cations for dimensional stability.

HYPOCHOLESTEROLEMIC EFFECT OF CHITOSAN IN HUMANS

Maezaki et al. (1993) published data on the dietary effects of chitosan in adult males. The chitosan in this study was administered in the form of biscuits over a study period of 4 weeks (no chitosan, chitosan, chitosan, and no chitosan). When chitosan was given in the diet (3–6 g/day), the serum total cholesterol level significantly decreased, whereas the serum high-density lipoprotein (HDL)-cholesterol level significantly increased when compared with the level for each of them before ingestion. This resulted in a significant decrease in the atherogenic index. Similar results were obtained by Sugano et al. (1988) with rats raised for 20 days after dietary administration of chitosan. However, the dose of chitosan in their study corresponds to as much as a few grams per day per kg of rat body weight, and in the same report, it was shown that 81-day feeding with a 0.5% chitosan-added noncholesterol diet did not induce any change in the serum cholesterol level. The results of Maezaki et al. (1993) on humans show a favorable effect with a very low dose within a short period and are thus worthy of special attention.

The chitosan intake (biscuits) by healthy volunteers was found by Terada et al. (1995) to produce a significant decrease of the fecal phenols, p-cresol and indole, in analogy with other polysaccharides. Chitosan inhibited the putrefactive activity of the intestinal microbiota, thus reducing the risks of disease states. The lecithinase-negative clostridia were the only microbiota significantly depressed by chitosan in humans, after a 2-week chitosan intake period.

Ventura (1996) and Veneroni et al. (1996) have conducted different experiments to determine the effect of a new chitosan dietary integrator in obese patients. Subjects were divided into two groups: one was treated with hypocaloric diet and chitosan dietary fiber, the other was treated with hypocaloric diet and a placebo for four weeks. At the end of the study period, a statistically significant reduction of the body weight and overweight, triglycerides, and total LDL cholesterol were observed in both groups, but in the chitosan treated group the differences were statistically greater than in the placebo group. From the results obtained in these studies, the diet plus a chitosan dietary fiber appears to be a useful treatment of the overweight and hyperlipidemia in obese subjects.

PERSPECTIVES FOR THE TREATMENT OF CELIAC DISEASE

Celiac disease is an enteropathy occurring in genetically predisposed individuals, when wheat is a component of the diet (Catassi et al.,

1994). Clinical studies have shown that rye, barley, and probably oat, in addition to wheat, are also toxic (Baker and Read, 1976; Weijers and Van de Kamer, 1955), whereas rice and maize are nontoxic and are used safely in the celiac patient's diet. Toxicity of cereals other than wheat in celiac disease is most likely due to prolamine fractions equivalent to gliadins (Auricchio et al., 1984; Pittschieler et al., 1994). The total peptic-triptic digest (PT-digest) of prolamine peptides from cereals toxic in celiac disease agglutinates K562(S) cells and inhibits the development of the small intestine of 17-day-old rat fetus and the improvement of cultured jejunal flat mucosa of celiacs (Auricchio et al., 1986). These biological activities, probably related to the noxious effects of these peptides for the celiac intestine, are all prevented by mannan (Auricchio et al., 1990) and oligomers of *N*-acetylglucosamine (Auricchio et al., 1987), suggesting that the agglutinating and toxic peptides are bound by these carbohydrates. Mannan or *N*-acetylglucosamine oligomers are able to bind all the peptides of a gliadin digest that agglutinate K562(S) cells and damage the fetal rat intestine in vitro (Auricchio et al., 1993). Methylpyrrolidinone chitosan (MP-chitosan) coupled to Sepharose-6B was used to isolate hexaploid wheat peptides endowed with the agglutinating activity by De Vincenzi et al. (1993).

5-Methylpyrrolidinone chitosan was synthesized according to Muzzarelli (1992) and Muzzarelli et al. (1993). The viscosity of the 1.0% chitosan solution was 13 mPa/s, instead of 48 mPa/s for the control (unbleached). The identity of chitin and its derivatives was monitored by infrared spectrometry. The coupling of 5-methylpyrrolidinone chitosan to the Sepharose-6B gel was performed as recommended by Pharmacia (Uppsala, Sweden) with a slight modification. The reaction mixture was coupled at 37°C for 48 h, and then the gel was charged with 0.1 M ethanolamine at pH 8 and kept at 4°C overnight. The coupling efficiency was found to be 57%.

Enzymatic digests of the prolamine fractions from wheat (*Triticum aestivum,* var. S. Pastore), rye (*Secale cereale,* var 500-2G), barley (*Hordeum vulgare,* var. Arma), and oat (*Avena sativa,* var. Astra) were prepared according to De Ritis et al. (1979). Each digest (~50 mg) was percolated at the flow rate of 24 mL/h through a methylpyrrolidinone chitosan-Sepharose-6B column (5 × 35 cm) equilibrated with 0.02 M ammonium acetate buffer, pH 7.2. Fractions A and B were collected, and the column was washed with the above buffer until no absorbance at 278 nm was detected in the effluent. Fraction C was eluted from the column with 0.1 N acetic acid (introduced at eluant/bed volume ratio 1.6), after the pH value dropped to 2.87. Fraction C was immediately neutralized by addition of 0.5 M ammonium hydroxide (~450 mL/tube). The yields are reported in Table 1.

Table 1. Fractions obtained from PT-digests of cereal prolamine submitted to chromatography on MP-chitosan, expressed as percent of the total loaded prolamine PT-digest.

Cereal	Fractions, %			
	A	B	C	Total
Bread wheat	1.5	92.0	1.30	94.80
Rye	0.8	90.0	1.10	91.90
Barley	1.1	91.5	1.26	93.36
Oat	1.2	90.0	1.00	92.20

The agglutination test was based on the use of K562(S) cells resuspended with Dulbecco phosphate buffer saline at a final concentration of 10^8 cells/ml. The PT-digests, under experimental conditions simulating the in vitro protein digestion, were very active in agglutinating undifferentiated K562(S) cells. Fractions A and B tested at the concentration of 14 g/L of culture medium were not active in agglutinating K562(S) cells; on the contrary, fraction C (~ 1.0–1.3% of the total prolamine loaded), showed higher agglutinating activity than the PT-digest itself. The minimum concentration required to agglutinate the totality of the cells in suspension was 1.6–25.0 mg/L of culture medium (Table 2).

The MP-chitosan was an effective inhibitor of the agglutination induced by PT-digest from bread wheat gliadin: in the presence of 2.333 g of bread wheat gliadin peptides per liter of cell suspension, MP-chitosan exerted 100% inhibition at the concentration of 2.3 g/L (Table 3). When added to agglutinated cells, MP-chitosan was able to dissociate them: in practice, the cell layer formed after 20 minutes in the presence of wheat gliadin peptides, disappeared under the effect of

Table 2. Agglutination of K562(S) cells by fractions of prolamine peptides from cereals, separated by chromatography on MP-chitosan. The minimal agglutinating concentration (MAC) is the minimal concentration in the agglutination test suspension required to agglutinate 100% of cells. For fractions A and B the highest concentration tested was 14 g/L.

Cereals	MAC, mg/L Total PT-Digest	A	B	C
Bread wheat	73	no	no	1.6
Rye	150	no	no	7.8
Barley	180	no	no	11.4
Oat	104	no	no	25.0

Table 3. Inhibitory effect of MP-chitosan on the agglutinating activity of gliadin peptides from bread wheat (2.333 g per liter of cell suspension).

MP-Chitosan Concentration, g/L	Agglutinating Activity, %
2.333	No agglutination
1.166	20
0.291	50
0.146	100

added MP-chitosan at the above mentioned concentration, and the cells regained their normal appearance.

The MP-chitosan was a polydisperse sample as shown by gel electrophoresis, with average MW in the 10^5 order. ^{1}H NMR spectrometry did not provide evidence of extended depolymerization (Muzzarelli et al., 1993). It had the advantage of being easily coupled to Sepharose at alkaline pH values. It is known that plants synthesize a wide array of proteins capable of binding to affinity matrices based on chitin. Those proteins contain one or more chitin-binding domains whose affinity is not restricted to chitin but may extend to complex glycoconjugates. Therefore, certain modified chitosans may prove even more effective than chitin in binding to lectins from *Triticum aestivum, Hordeum vulgare, Oryza sativa,* and *Secale cereale*, which share similarities according to immunological, biochemical, and carbohydrate-binding criteria (Raikhel et al., 1993).

The evidence gained at the analytical level indicates that perspectives exist for the exploitation of the selective interaction of gliadin and MP-chitosan, in view of the prevention of the celiac disease.

HYPOURICEMIC EFFECT OF CHITIN AND CHITOSAN

Chitin and chitosan are used in health food for the prevention and treatment of hyperuricemia. Male rats were fed with a feed supplemented with cellulose (control), chitin, or chitosan, and the uric acid in blood and urine was measured. The animals fed with the addition of chitosan had much lower uric acid in blood and urine than those of control (Maekawa and Wada, 1990).

Rats were administered adenine (control) and chitosan together with adenine. Although the uric acid concentration increased fourfold within 24 days in controls, no increase was reported for the chitosan-treated animals (Mita et al., 1989).

The effect of dietary fiber intake on the digetsion of nucleic acids, the inhibition of hyperuricemia by chitosan, and the effect of chitosan intake on the metabolism of uric acid were reviewed by Wada (1995).

CONCLUSIVE REMARKS

The oral administration of chitosan has posed the following questions: 1) consequences on the amount of iron in the body; 2) influence on the intestinal flora; 3) effect on lipases; and 4) most suitable forms for administration. All of them have been addressed by various authors.

Jennings et al. (1988) suggested that chitosan in rats does not reduce serum iron or hemoglobin and has lipid-lowering effects equivalent to those of cholestyramine without producing similar deleterious effects on intestinal mucosa. The chelating ability of chitosan could in fact represent a worry in terms of depletion of iron. This subject was also addressed by Gordon and Besch-Williford (1983) with similar conclusions. Breads containing chitosan-treated heme iron were also proposed by Chiba (1988).

A burned mouse model was set up by Nelson et al. (1994) to study the influence of dietary fiber on bacterial translocation after burn injury. The animals that received chitosan had significantly lower levels of bacteria in the cecum, mesenteric lymph nodes, and liver; thus, chitosan helped in reducing the bacterial translocation.

The lipase and lipoprotein lipase activities are of interest in the present context. In fact, if chitosan is a substrate for human lipase, it might be possible that the relatively large amount of chitosan introduced with the diet would in part prevent lipases from hydrolyzing the lipids. It is well known that 2-monoglycerides and fatty acids are the partial hydrolysis products, yielded by pancreatic lipase, suitable for absorption. Studies on the lipase-chitosan system are in progress and include wheat germ lipase (Muzzarelli et al., 1995), porcine pancreatic lipase, and microbial lipase (Pantaleone et al., 1992). Moreover, chitosan could form salts with bile acids and substract them from the emulsification process.

The chitosan intended for oral administration should preferably not be submitted to thermal treatment such as cooking and baking. Although biscuit cooking imparts appealing organoleptic properties, the Maillard reaction would partly destroy the primary amino group of chitosan and reduce the efficacy as a dietary fiber (Tanaka et al., 1993). This fact should be taken into account when baked products are proposed.

The basicity of the amino group, which in early times directed attention to the antiulcer activity of chitosan (Doczi et al., 1964; Hillyard et al., 1964; Weisber and Gubner, 1966), is the key factor for the biological significance of this dietary fiber, which is expected to be useful in the *per os* treatment of arthrosis and osteoporosis as well (Ito, 1995).

ACKNOWLEDGEMENT

The skillful assistance of Mrs. Maria Weckx in retrieving the bibliographic material is gratefully acknowledged. Work supported with Fondi Quaranta Percento MURST.

REFERENCES

Artursson, P., T. Lindmark, S. S. Davis and L. Illum. 1994. "Effect of Chitosan on the Permeability of Monolayers of Intestinal Epithelial Cells (Caco-21)," *Pharm. Res.,* 11(9):1358–1361.

Auricchio, S., G. De Ritis, M. De Vincenzi, F. Latte and V. Silano. 1984. "An in vitro Animal Model for the Study of Cereal Components Toxic in Coeliac Disease, *Ped. Res.,* 18:1372–1378.

Auricchio, S., G. De Ritis, M. De Vincenzi, M. Maiuri and V. Silano. 1987. "Prevention by Mannan and Other Sugars of in vitro Damage of Rat Fetal Small Intestine Induced by Cereal Prolamine Peptides Toxic for Human Celiac Intestine," *Ped. Res.,* 18:1372–1378.

Auricchio, S., G. De Ritis, M. De Vincenzi and V. Silano. 1986. "Toxicity Mechanisms of Wheat and Other Cereals in Celiac Disease and Related Enteropathies," *J. Ped. Gastr. Nutr.,* 4:923–930.

Auricchio, S., G. De Ritis, M. De Vincenzi, E. Mancini, M. Maiuri and V. Silano. 1990. "Mannan and Oligomers of *N*-Acetylglucosamine Protect Intestinal Mucosa of Coeliac Patients with Active Disease from in vitro Toxicity of Gliadin Peptides," *Gastroenterol.,* 99:973–978.

Auricchio, S., G. De Ritis, M. R. Dessì, M. De Vincenzi, M. Maiuri and E. Mancini. 1993. "Selective Binding to Mannan of Gliadin Peptides Which Agglutinate Undifferentiated K562(S) Cells and Inhibit in vitro Development of Fetal Rat Intestine," *Gastroenterol.,* 104:233.

Baker, P. G. and A. E. Read. 1976. "Oat and Barley Toxicity in Coeliac Patients," *Postgrad. Med. J.,* 52:264–267.

Catassi, C., I. M. Ratsch, E. Fabiani, M. Rossini, F. Bordicchia, F. Candela, G. V. Coppa and P. L. Giorgi. 1994. "Coeliac Disease in the Year 2000: Exploring the Iceberg," *Lancet,* 343:200–203.

Chiba, Y. 1988. "Manufacture of Breads Containing Chitosan-Treated Heme Iron," *Jpn. Kokai Tokkyo Koho,* JP 01,179,638.

De Ritis, G., P. Occorsio, S. Auricchio, F. Gramenzi and V. Silano. 1979. "Toxicity of Wheat Flour Proteins and Protein Derived Peptides for in vitro Developing Intestine from Rat Fetus," *Ped. Res.,* 13:1255–1261.

Deuchi, K., O. Kanauchi, Y. Imasato and E. Kobayashi. 1994. "Decreasing Effect of Chitosan on the Apparent Fat Digestibility by Rats Fed on a High-Fat Diet," *Biosc. Biotech. Biochem.,* 58(9):1613–1616.

De Vincenzi, M., M. R. Dessì and R. A. A. Muzzarelli. 1993. "Separation of Celiac Active Peptides from Bread Wheat with the Aid of Methylpyrrolidinone Chitosan," *Carbohydr. Pol.,* 21:295–298.

Doczi, J., F. C. Ninger and H. I. Silverman. 1964. "Composition for Inhibiting Pepsin Activity and Method of Preparing Same," U.S. Patent 3,155,575.

Fukada, Y., K. Kimura and Y. Ayaki. 1991. "Effect of Chitosan Feeding on Intestinal Bile Acid Metabolism in Rats," *Lipids,* 26(5):395–399.

Furda, I. 1980. "Nonabsorbable Lipid Binder," U.S. Patent 4,223,023.

Furda, I., ed. 1983. *Unconventional Sources of Dietary Fiber.* Washington, DC: Am Chem. Soc.

Furda, I. and C. J. Brine, eds. 1990. *New Developments in Dietary Fiber.* New York, NY: Plenum.

Gordon, D. and C. Besch-Williford. 1983. "Chitin and Chitosan: Influence on Element Absorption in Rats," in *Unconventional Sources of Dietary Fiber.* I. Furda, ed., Washington, DC: Am. Chem. Soc.

Hillyard, I. W., J. Doczi and P. Kiernan. 1964. "Antacid and Antiulcer Properties of Chitosan in the Rat," *Proc. Soc. Exp. Biol. Med.,* 115:1108–1112.

Ikeda, I., M. Sugano, K. Yoshida, E. Sasaki, Y. Iwamoto and K. Hatano. 1993. "Effects of Chitosan Hydrolysates on Lipid Absorption and on Serum and Liver Lipid Concentration in Rats," *J. Agr. Food Chem.,* 41:431–435.

Ito, F. 1995. "Role of Chitosan as a Supplementary Food for Osteoporosis," *Gekkan Fudo Kemikaru,* 11(2):39–44. CA 122:186221.

Jennings, C. D., K. Boleyn, S. R. Bridges, P. J. Wood and J. W. Anderson. 1988. "A Comparison of the Lipid-Lowering and Intestinal Morphological Effects of Cholestyramine, Chitosan and Oat Gum in Rats," *Proc. Soc. Exp. Biol. Med.,* 189:13–20.

Kanauchi, O., K. Deguchi, Y. Imasato and E. Kobayashi. 1994. "Increasing Effect of a Chitosan and Ascorbic Acid Mixture on Fecal Dietary Fat Excretion," *Biosc. Biotech. Biochem.,* 58(9):1617–1620.

Maekawa, A. and M. Wada. 1990. "Food Containing Chitin or Its Derivatives for Reduction of Blood and Urine Uric Acid," *Jpn. Kokai Tokkyo Koho,* JP 03,280,852. CA 116:127402.

Maezaki, Y., K. Tsuji, Y. Nakagawa, et al. 1993. "Hypocholesterolemic Effect of Chitosan in Adult Males," *Biosc. Biotech. Biochem.,* 57(9):1439–1444.

Mita, N., T. Asano and K. Mizuochi. 1989. "Pharmaceuticals or Food Containing Edible Fibers of Chitosan for Hyperuricemia Control," *Jpn. Kokai Tokkyo Koho,* JP 02,311,421. CA 114:150221.

Muzzarelli, R. A. A. 1992. "Depolymerization of Methyl Pyrrolidinone Chitosan by Lysozyme," *Carbohydr. Polym.,* 19:29–34.

Muzzarelli, R. A. A., P. Ilari and M. Tomasetti. 1993. "Preparation and Characteristic Properties of 5-Methyl Pyrrolidinone Chitosan," *Carbohydr. Polym.,* 20:99–106.

Muzzarelli, R. A. A., W. Xia, M. Tomasetti and P. Ilari. 1995. "Depolymerization of Chitosan and Substituted Chitosans with the Aid of a Wheat Germ Lipase Preparation," *Enz. Micr. Technol.,* 17:541–545.

Nauss, J. L., J. L. Thompson and J. Nagyvary. 1983. "Hypocholesterolemic Effect," *Lipids,* 18:714–719.

Nelson, J. F., J. Wesley Alexander, L. Gianotti, C. L. Chalck and Pylesw. 1994. "Influence of Dietary Fiber on Microbial Growth in vitro and Bacterial Translocation after Burn Injury in Mice," *Nutrition,* 10(1):32–36.

Pantaleone, D., M. Yalpani and M. Scollar. 1992. "Unusual Susceptibility of Chitosan to Enzymic Hydrolysis," *Carb. Res.,* 237:325–332.

Pittschieler, K., B. Ladinser and K. Petell. 1994. "Reactivity of Gliadin and Lectins with Celiac Intestinal Mucosa," *Ped. Res.,* 36(5):635–641.

Pupita, F. and A. Barone. 1987. "Action of DEAE-D on Lipid Digestion and Absorption," *Int. J. Obesity,* 11(3):13–23.

Raikhel, N. V., H. I. Lee and W. F. Broekaert. 1993. "Structure and Function of Chitin-Binding Proteins," *Annu. Rev. Plant. Physiol. Plant. Mol. Biol.,* 44:591–615.

Sugano, M., S. Watanabe, A. Kishi, A. Izume and A. Ohtakara. 1988. "Hypocholesterolemic Action of Chitosans with Different Viscosities in Rats," *Lipids,* 23(3):187–191.

Sugano, M., K. Yoshida, M. Hashimoto, K. Enomoto and S. Hirano. 1992. "Hypocholesterolemic Activity of Partially Hydrolyzed Chitosans in Rats," in *Advances in Chitin and Chitosan,* C. Brine, J. P. Zikakis and P. Sandford, eds., Amsterdam: Elsevier.

Tanaka, M., J. R. Huang, W. K. Chiu, S. Ishizaki and T. Taguchi. 1993. "Effect of the Maillard Reaction on Functional Properties of Chitosan," *Nippon Suisan Gakkaishi,* 59(11):1915–1921.

Terada, A., H. Hara and D. Sato, et al. 1995. "Effect of Dietary Chitosan on Faecal Microbiota and Faecal Metabolites of Humans," *Microb. Ecol. Health Dis.,* 8:15–21.

Tsugita, T. and K. Sakamoto. 1995. "Cholesterol Improvement Effect of Chitosan and Its Application to Food," *Gekkan Fudo Kemikaru,* 11(2):45–50. CA 122:158823.

Vahouny, G. V., S. Satchihanandam, M. M. Cassidy, F. B. Lightfoot and I. Furda. 1983. *Am. J. Clin. Nutr.,* 38:278–284.

Veneroni, G., F. Veneroni, S. Contos, S. Tripodi, M. De Bernardi, C. Guarino and M. Marletta. 1996. "Effects of a New Chitosan on Hyperlipidemia and Overweight in Obese Patients," in *Chitin Enzymology, Vol. 2,* R. A. A. Muzzarelli, ed., Ancona, Italy: Atec Edizioni.

Ventura, P. 1996. "Lipid Lowering Activity of Chitosan, A New Dietary Integrator," in *Chitin Enzymology, Vol. 2,* R. A. A. Muzzarelli, ed., Ancona, Italy: Atec Edizioni.

Wada, M. 1995. "Effect of Chitin, Chitosan Intake on Metabolism of Uric Acid," *Gekkan Fudo Kemikaru,* 11(2):25–31. CA 122:186062.

Weijers, H. A. and J. H. Van de Kamer. 1995. "Experiments on Cause of Harmful Effect of Wheat Gliadin," *Acta Pediatr.,* 44:465–469.

Weisber, M. and R. S. Gubner. 1966. "Chitosan Containing Antacid Composition and Method of Using Same," U.S. Patent 3,257,275.

CHAPTER 8

Enhancing Food Production with Chitosan Seed-Coating Technology

DONALD FREEPONS
P.O. Box 7134
Kennewick, WA

INTRODUCTION

The world's food supply is produced from a finite area of land. With ever-increasing populations creating greater demands on our food supplies, new crop production methods are urgently needed. Moreover, this acute need for increasing food production results not only as the world's population grows, taxing the soils for adequate food supplies, but as food-producing areas are shrinking because of the continuous urban sprawl. Add to this, the loss of crop lands to erosion, misuse, and contamination. In other words, we need more food produced from fewer acres.

Cropping soil produces the basic supply of food. This top soil (or cropping soil) is, by comparison, a microscopic film or layer, of the surface of our planet's crust. In reality, the productive soil layer is, on average, a 2-foot-thick layer on the earth's 20-mile-thick crust, and yet it is one of the main providers of food. Almost all life is dependent on the production of this top soil. It is this soil that fed early man. The animals that inhabit the earth are dependent on crop soils. All life is held in the balance of food production versus food consumption. Yet, interestingly, the extraordinarily broad field of biology provides uniquely adapted plants that can reproduce in *almost every climate*! In arid areas, grasses feed livestock. During periods of annual drought, these plants have the ability to become dormant. With rains, they begin growth and provide the needed forage for animals. A wide variety of plants are tolerant of different amounts of rainfall, from the millets in very low rainfall areas to rice that thrives in flooded soils. Likewise, grains and tuber crops produce abundantly in colder regions, whereas bananas and mangos require high growing temperatures. In summary, there are special plants that can accommodate each increment in wide ranges of moistures and temperatures and are distributed accordingly to provide food for the world.

It is with these basic food crops and their related growing requirements that agricultural scientists find their work; their challenge is to produce abundant, nutritionally safe, and economical food supplies for the world's growing populations.

Science and industry have contributed much to this equation. Early on, the soil was tilled to increase population. And then, plant selection was recognized as a means of increasing crop yields. Growers undoubtedly noticed some plants in their fields were superior to the general plant population. These plants were then selected and their seeds harvested separately, saved, and used for planting stock the next crop season to improve the crop.

From this humble beginning, agroscience began. Probably during this same time, fertilizers made their appearance. With large numbers of domestic animals on every farm, the disposal of the manure became a problem. Perhaps some clever farmer noticed that when he spread it onto his fields, the crops grew better, thus, the beginning of fertilization. Then, phosphates were introduced to immobilize the ammonia from aerating manure piles. Next, came the basic nutrients, nitrogen, phosphorus, and potassium, in various combinations and forms. Trace elements entered the arsenal of plant nutrients as production tools. It was a grand time in crop production with each new entity increasing the crop yields.

The last and most recent members of the new food-producing technologies were the pesticides. With their advent, farmers could control insects, weeds, and disease. Crop yields exploded! We had entered the "Green Revolution." As these technologies dovetailed to increase yields, it became apparent to the producers that they had reached the peak of their land's production. Split fertilizer applications along with computerized cropping management inched up yields, but they could go only so far. Overseeding and overfertilizing actually reduced yields. And environmentalists, along with water and air quality agencies, have complained of agrochemical residues in aquifers and in foods. Their concern included pesticides that remained air-borne far from their point of application. Have we reached the limit in food production? Are we in effect, poisoning our planet with overapplications of agrochemicals?

Can science find new ways to increase food production from a depleting area? A search began for a different mode, e.g., a biochemical system to increase crop production.

The key task for agricultural and biological research scientists is to develop new techniques that can provide an increasing and abundant supply of nutritious and safe foods that are affordable to the masses. We need new tools to enhance food production. Chitosan seed-treating technology, a leader in these new substances, consistently increases crop yields.

HISTORY OF PHYTOHORMONES

In the 1930s and 1940s, a small group of natural products was discovered. It was inconceivable then that plant hormones other than indole-3-acetic acid (IAA), gibberellins (GA_n), and ethylene would be discovered. But it was being noticed that these materials could not explain findings from experimental and practical plant physiology [2]. Plant scientists proceeded to look for other hormones. Cytokinins and abscisic acid were discovered and added to the supposedly complete list of the five endogenous plant hormones:

1. Auxins
2. Cytokinins
3. Gibberellic acids
4. Abscisic acid
5. Ethylene

These hormones are pleiotropic, rather than specific, that is, each has more than one effect on the growth and development of plants having a complex array of functions [3]. These substances have a wide range of influences on plant development, seemingly from one extreme to the other, e.g., from inhibiting growth to stimulating growth, depending on the genus and species and the stage of development. As bioscientists explored and manipulated these substances experimentally, there still were unanswered questions pertaining to plant development.

Results pointed to substances that had a high degree of specificity.

There seemed to be a regimen of natural substances that are a class above the known hormones. Albersheim et al. found that a small subset of genes as a complex system of chemical messengers could cause cell differentiation [3]. Complex carbohydrates as cell wall fragments could lay dormant or latent waiting for a certain stage of plant development and for an enzyme system to activate them. Perhaps this was the entry of chitin and chitosan and their derivative molecules to the new order of plant hormone system:

1. Oligosaccharides–(chitosan)
2. Polyamines
3. Brassinosteroids
4. Salicylic acid
5. Tricontinol
6. Methanol
7. Vitamins

If we looked at all of these substances, singularly or in combination, they do answer and explain the enigma of hormonally regulated plant growth and development. When the information is dovetailed with the known plant growth regulators, something is still missing, implying further interactions with other cofactors and, perhaps as yet, undiscovered hormones [2]. Chitosan seed-implanting technology represents this new "tier" in the hierarchy of biological plant controls and influences. It is the forerunner and pacesetter to these newly discovered bioregulators that cannot be explained by the five well-known hormones.

CHITOSAN BIOCHEMISTRY

Chitosan's bioactivity comes from a complex amine molecule that carries a positive charge. Chitosan, the partially deacetylated form of chitin, is a natural biopolymer composed of the two common sugars, glucosamine (2-amino-2-deoxy-D-glucose, ~70–80%) and N-acetylglucosamine (2-acetamido-2-deoxy-D-glucose, ~20–30%) [4]. It is a member of a class of saccharides called amino sugars and shares this class of sugars with galactosamine and mannosamine, all very active biologically. They are structural components in exoskeletal animals (arthropoda) and fungi. Two other amino sugars, murimic and neuraminic acids, are building blocks found in cell walls of bacteria and cell coats of higher animal cells. Moreover, amino sugars characteristically have unique viscosities and, as such, are members of a subclass of sugars called mucosaccharides. Biologically, they are building blocks or components of tendons, the cornea, and vitreous humor. They also exhibit unusual properties associated with their lubricities and are components, along with hyaluronic acid, in the synovial fluids. In addition, science has found that when the chitosan solution dehydrates, it forms a unique semipermeable membrane or film that has found usage in pharmaceuticals and cosmetics [4].

Early on, it became evident that the chitosan molecule is fragile. This unusual attribute posed certain problems, causing early attempts to commercialize chitosan seed coating to fail. For example, it was found that a chitosan solution has a finite shelf life. The source of the chitin is a factor in chitosan's bioactivity. Moreover, some solubilizers are phytotoxic, thus they can negate bioactivity in seed treating. Chitosan's unique viscosity index presented problems in coating seeds. Available equipment was unable to pump and atomize the viscous fluids.

As each problem was recognized and addressed, the successes showed the dependability and consistent performance of a chitosan seed coating. Further, advantages were found in the viscous chitosan solution in that other agents can be encapsulated integrally in this soon-to-form film. For example, fungicides could be tenaciously attached to seeds to resist wash off when planted through paddy conditions. Further, there was no contamination of the paddy water from escaping chemicals.

CHITOSAN BENEFITS

We have learned and documented that chitosan through seed implantation can regulate, direct development, and manipulate the unfolding of the plant's growth. Plant stands are improved by chitosan seed coatings. Healthier germinating seedlings improve the plant population. These higher plant populations effectively reduce weed pressures. Higher plant populations in the field also support and contribute to higher yields.

The first requirement to produce a bountiful harvest is an adequate plant stand in the field. How can a producer be assured of an adequate plant stand in the producer's field? The plant stand is the foundation for a bountiful crop whether it is rice, wheat, peppermint, potatoes, asparagus, or any crop. Chitosan seed treating has increased seedling populations up to 25% greater than the untreated control fields. As a result, commercial rice seeding rates have been reduced because of overpopulation as a result of the influence. Root development was increased by chitosan seed treatments. Winter wheat seedlings 68 days after planting produced 14% greater dry mass (predominant root matter) compared with their counterpart, the plants grown from untreated seeds.

Rice plants grown from seed treated with chitosan exhibited increased tillering (7.7%) and better pollination (head weight, 21.3% at milk stage) compared with the untreated control.

Physiologically, plants grown from seeds that have been implanted with chitosan are more productive compared with plants grown from untreated seeds. For example, rice and wheat fields planted with chitosan-treated seed, consistently produce from 5% to 20% more harvested grain. Consistently, commercial rice fields have outproduced

Table 1. Field trial no. 1. Chitosan:
Louisiana rice field trials.

Planting Data	Control	Treatment
Area	6.9 ha	4.0 ha
Seeding Rate	168.0 kg/ha	158.8 kg/ha
Fertilizer Rate	130.0 kg/ha	130.0 kg/ha
Yield Data		
	6172.5 kg/ha	7110.6 kg/ha
Increased yield from chitosan seed implant		938 kg/ha

the untreated controls by 450–1,000 kg/ha under equal conditions. Potato seed pieces that received chitosan seed treatment produced 11.8% greater tonnage per acre over their counterpart, and the quality (total solids measured by the specific gravities) was similarly improved.

These chitosan-influenced yields have been established from 5 years of university research trials to provide a solid database for the chitosan seed-coating technology. Moreover, the chitosan seed-coating technology has been transferred to large commercial acreage. From five harvest years under two different rice-growing regions, these fields produced data that substantiated the university findings.

A typical example of chitosan influence on *drill-seeded rice* is shown in Table 1. A typical example of the chitosan influence on *water-seeded rice* is shown in Table 2. Table 3 is an example of the chitosan influence on *water-seeded rice* where *red rice as a weed, caused a "serious problem."* The chitosan treatment outyielded the control. The important differences pertain to the increased yield under equal fertilizer application rates and, because the seeding rate was reduced, economic benefits.

Table 2. Field trial no. 2. Chitosan:
Louisiana rice field trials.

Planting Data	Control	Treatment
Area	5.26 ha	5.78 ha
Seeding Rate	162.9 kg/ha	147.9 kg/ha
Fertilizer Rate	128.0 kg/ha	128.0 kg/ha
Yield Data		
	4604.0 kg/ha	5629.0 kg/ha
Increased yield from chitosan seed implant		1025.0 kg/ha

Table 3. Field trial no. 3. Chitosan:
Louisiana rice field trials.

Planting Data	Control	Treatment
Area	9.3 ha	7.1 ha
Seeding Rate	201.0 kg/ha[a]	145.0 kg/ha
Fertilizer Rate	145.0 kg/ha	145.0 kg/ha
	Yield Data	
	3928.3 kg/ha	4145.6 kg/ha
Increased yield from chitosan seed implant		215.2 kg/ha

[a]This grower opted to top-seed with an additional 56 kg/ha.

An example of rice grown under the *continuous flood system* is demonstrated in Table 4. These yield increases fall within a consistent range, but the growing conditions are quite different. The California trial was a relatively new soil, i.e., less than 50 years of cropping, whereas the Louisiana trial was on land with more than 150 years of continuous cropping. Moreover, the California climate is semiarid, whereas Louisiana has very high humidity, causing high-disease pressures. Yet the chitosan seed treatment produced comparable yield increases (see also Table 5). Plants grown from wheat seeds treated with chitosan exhibit similar yield enhancements (Tables 6–8).

MODE OF ACTION

The literature is burdened with proposals of chitosan enhancing the plant's immune system [3,5]. The literature cites many proposals in which a hexosamine sugar could function as an elicitor to initiate the synthesis of phytoalexins [5]. Using spacer-filling molecule models, it is interesting to notice that an "elicitor," a sugar residue, when fully

Table 4. 1993 Chitosan: California rice field trials.

Planting Data	Control	Treatment
Area	4.5 ha	16.2 ha
Fertilizer Rate	174.0 kg/ha	174.0 kg/ha
	Yield Data	
	7860.2 kg/ha	8945.4 kg/ha
Increased yield from chitosan seed implant		1085.3 kg/ha

Table 5. University replicated rice trials.

	Control	Treatment	Yield Increase
Trial No. 1	3019.5	3269.5 kg[a]	250.0 kg
Trial No. 2	1918.8	2058.5 kg	139.7 kg
Trial No. 3	2547.6	2892.0 kg	344.8 kg
Trial No. 4	2620.0	2924.7 kg	303.9 kg
Trial No. 5	2620.0	3034.0 kg	413.3 kg

[a]Weights represent total plot weights calculated for kg/ha basis.

Table 6. Trial no. 1. California dryland wheat trial.

Control	Treatment	Yield Improvement
2293.8 kg/ha	3075.5 kg/ha	+781.8 kg/ha

Table 7. Trial no. 2. California irrigated wheat trial.

Control	Treatment	Yield Improvement
6424.3 kg/ha	7056.0 kg/ha	+631.7 kg/ha

Table 8. 1993 Louisiana rice plant count data:
plant population (average plants/ft²).

	Control	Treatment	Samples Taken
Field No. 1	14.0	17.7	$N = 22$
Field No. 2	15.5	20.4	$N = 13$
Field No. 3	18.6	23.3	$N = 23$
Field No. 4	17.8	22.1	$N = 14$

rotated, can uniquely fit into the DNA double helix, hydrogen bonding to the phosphate oxygens of DNA [6].

Perhaps chitosan as a cell-wall oligosaccharide fragment can function as a latent or dormant messenger, waiting to give the cell directions in which to develop. Cell-wall fragments can "turn on" the cell at a given moment [3].

There seem to be cues or nuances that indicate chitosan has a specific bioactivity. It is biologically possible that the chitosan molecule is a true plant growth regulator and functions as a gene expression. With documented increased root development, which would have greater capabilities to scavenge nutrients, why doesn't the foliage appear darker green, indicating additional nitrogen is being delivered to the plant? Moreover, plants from chitosan-treated seeds are not greater in height. Therefore, chitosan can be a complex glucoprotein that specifically cites the reproductive phase of the plant's development, e.g., tillering and pollination. Chitosan appears to have multiple influences directed toward the improvement of the reproductive phase. Much is yet to be discovered. At present, chitosan's consistent performance as a yield enhancer is reliable.

Chitosan films provide an ancillary or secondary benefit. This unique film has shown utility in delivering fungicides with seed coating because it forms a semipermeable membrane or film that encompasses each seed in a cellophane-like encasement.

It has been documented that this film can effectively encapsulate fungicides that control soil and seed-borne diseases. Fungicides are integral within this film, thus they cannot be buffed or chaffed from the seed. In treating preplant seed potato pieces, the retention of fungicides, e.g., Tops™ is increased markedly. Second, this film, upon hydration, forms a gelatinous shroud entombing the fungicide within. Any invading organism confronts this gelatinous barrier before it can contact the seed.

Under conditions unique to direct-seeded rice (paddy seeding), environmental safety is maintained because this barrier holds the fungicide tenaciously to the seed in a undilutable and nonerodible state, sealed against water contamination.

Likewise, with foliar-applied pesticides, after an agrochemical has been bound into the chitosan system and applied and deposited on a leaf surface, it is tenaciously attached, thus resisting removal by wind or rain. The pesticides are integral within this film where they are intimately available to the targeted organism over an extended time frame. These attributes of the chitosan film maintain the pesticidal activity because no material is lost. As a result, pesticide application rates can be reduced. In its hydrated form, chitsoan film has a gelatinous composition with the pesticide encapsulated within, available to be metered into the plant system. Chitosan film is environmentally safe and an effective carrier for delivering agrochemicals. This chitosan film could pioneer the development of controlled delivery systems for agrochemicals.

Science and the food industry are recognizing the importance of controlled delivery systems in which the pesticide is maintained and regulated on the leaf surface to increase efficacy and to provide zero contamination to the environment.

SUMMARY

Chitosan, a complex and naturally occurring molecule, has improved a food plant's development, resulting in increased (grain) production. Chitosan is the pacesetter for this new bioactive overlay of hormone and enzyme systems. Just as fertilizer chemicals can increase the growth of a plant, these new hormone-based molecules can specifically be targeted to focus on one system of plant physiology to regulate a particular phase of its development.

Presently, yields are tied to the fertilizer input; an increase in fertilizer applications increases yields. Conversely, a reduction in fertilizer application reduces yields. But, there is a limit to the amount of fertilizer that can increase yields. Overapplication of fertilizers produce an abundance of foliage without increasing seed set. Heavy foliage is vulnerable to disease. Consequently, ultrahigh fertilizer rates, in reality, reduce yields. How then, can yields be increased? Chitosan technology influences the physiology of the plant toward higher yields. Disease is reduced from enhanced immunity. The plant's physiology is controlled to produce grain rather than increased foliar growth.

Since there is a critical need for research that improves plant production while reducing the dependency on fertilizers, our research work has been directed toward reduced input agriculture. Is chitosan chemistry a means to increase food supply? Yes! We can look forward to more effective and increased food production from fewer acres in addition to protecting the environment.

REFERENCES

1. Freepons, D. E. March 14, 1989, U.S. Patent 4,812,159; Freepons, D. E. October 23, 1990, U.S. Patent 4,964,894.
2. Cutler, H. G. *Brassinosteroids Chemistry, Bioactivity and Applications.* American Chemical Society.
3. Albersheim, P. *Scientific American,* September, 1985.

4. Sandford, P. *Frontiers in Carbohydrate Chemistry.* 1990. Unpublished.
5. Hadwiger, L. Department of Plant Pathology. Washington State University.
6. Wither, F. H. and Gustine, D. L. Pennsylvania State University, American Chemical Society. 1991.

CHAPTER 9

Inhibition of Molting in Chewing Insect Pests

MARIA L. BADE

Center for Cancer Research
Massachusetts Institute of Technology
Cambridge, MA 02139

Humankind's struggle to protect itself from ravages inflicted by insect pests has a long history. However, insects tend to win in this ongoing combat because they are even more adaptable than humans and have much shorter generation times. Humans may gain a momentary respite from disease-transmitting flies by sending out millions armed with fly swatters, or temporarily gain the upper hand against body lice and mosquitoes by inventing DDT. But a safer long-term strategy lies in integrated pest management and in closer understanding of life cycle phenomena that may be vulnerable to constructive interference in which destructive pests are specifically targeted so that attacks on them are otherwise environmentally benign.

One underexplored event for interference of this nature is the molt in insects. All insects must undergo molts periodically to grow and complete their life cycle. The molt involves detachment of the about-to-be-outgrown cuticle from the underlying hypodermis (apolysis) and deposition of a new, in the larval stages potentially larger, cuticle beneath the old one, which is resorbed and recycled. These processes involve enzymes secreted into the space (ecdysial space) between old and new cuticle. Enzymes appear as zymogens in the ecdysial space and are apparently contained in two discrete sets of lysosomes. They are activated by a change in ionic environment followed by enzymatic processes and are at the last themselves resorbed. But unless it can be extensively degraded first, the cuticle cannot be "shed," which in *Manduca* larvae

141

involves gluing the claspers to the substrate on which they are molting, followed by motions in each segment and then a forward motion of the anterior end. Ultimately, the whole larva with its newly constituted new cuticle moves forward, whereas the now sheer old cuticle remnant, which includes the lining of the major tracheae and gut, is worked backward over the animal's surface and ultimately is left behind (Bade, manuscript in preparation).

In the caterpillar stage of Lepidoptera, cuticles consist of a rather thick chitin-containing endocuticle overlain by an exocuticle; outside of these is found a very thin, nonchitinous epicuticle. It has been known for a long time that chitin in intact endocuticle is not attacked by chitinases [1, H. Lipke, personal communication]. This seems to be caused by the fact that the faces of chitin subassemblies during the intermolt phase are masked by proteins that seem to a large part adsorbed to chitin, rather than covalently linked to it. David Smith showed with the aid of electron microscopy that the earliest stage of degradation of old cuticle in each molt consists in a visible loosening of the chitin-protein assemblies in it [1,2]. This permits the chief protease of molting fluid, an unusual metal chelator-sensitive one, to enter between microfibrils and clean off both adsorbed and covalently linked proteins. Bade and Shoukimas [3] showed that, in addition to the trypsin-like proteases previously reported, molting fluid possesses a proteolytic enzyme that is more active than the two trypsin-like ones, which show as a single sharp peak at pH 7.7, superimposed on the high-activity curve for the neutral protease that peaks at pH 7. It was shown that the trypsin-like proteases can be inhibited without interfering with activity of the chief protease, and that this indispensable enzyme is amidatic rather than esteratic, which is why its proteolytic activity will not be observed with an artificial protease substrate like TAME [4].

Bade and Shoukimas [3] studied the neutral chelator-sensitive protease in cacodylate, a buffer that does not contain phosphate, and with hemoglobin as the substrate. Enzyme activity was variable and seemed partially inhibited in the phosphate buffer used initially. A variety of inhibitors, which are listed in Table 1, were explored in cacodylate medium. As is apparent, ortho (or 1,10)-phenanthroline (OP) is a potent inhibitor for this protease, whereas some other inhibitors showed no effect. 8-Hydroxyquinoline-5-sulfonate (8-HQ), however, proved just as inhibitory *in vitro* as OP. (To improve solubility, the 5-sulfonate was used.)

Once *in vitro* inhibition with OP and 8-HQ was observed, extensive feeding experiments were initiated with a variety of chelators, to see which one(s) might inhibit molting when ingested along with the diet

Table 1. Effect of inhibitors on proteolytic activity in vitro.

Inhibitor	Residual Activity (%)
None	100 ± 8.2
Cysteine	104 ± 16.3
EDTA	25 ± 16.7
8-Hydroxyquinoline	27 ± 10.1
1,10-Phenanthroline	17 ± 8.6
p-Hydroxymercuribenzoate	105 ± 10.8
NaF	100 ± 13.3
Ovomucoid trypsin inhibitor	16 ± 10.2
Soybean trypsin inhibitor	22 ± 9.1

Reproduced with permission from Reference [5].

Preincubation: molting fluid in M/20 cacodylate pH 7 with 5×10^{-3} M final concentration of inhibitor (2.5 μg of trypsin inhibitors); 1 hr in ice.

Reaction: initiated with the addition of substrate (2 mg hemoglobin in 0.06 N HCl), incubated for 2 hr at 37°C in the presence of M/1,000 $CaCl_2$.

Analysis: 100 μl TCA-soluble supernatant; ninhydrin color read at 570 nm. The stated error is the sum of the percentage ranges for the mean zero time and experimental values. Three determinations per value shown.

[5]. It was expected that if a protease inhibitor was administered that inactivated the metal chelator-sensitive protease in vitro, chitinases might be inhibited as well, because as mentioned above, they can act only subsequently to the proteolytic removal of the proteins that invest chitin. *In vitro* active peptide protease inhibitors (Table 1) were not included in the feeding trials since it was felt that they would not survive passage through the midgut.

The expectation of inhibition of larval molts through blocking a crucial step in the process of molting, was met for *Manduca*. To compare a large number of measurements obtained in the feeding trials, a growth index was calculated for *Manduca*:

$$\text{Growth Index} = \frac{\text{Length (inches)} \times \text{Width (inches)}}{\text{Number of animals in group}} \times 100$$

Figure 1 shows OP-inhibited *Manduca* larvae next to the uninhibited controls. Both are the same age, i.e., the controls are near the end of the fourth larval instar. The small larvae, which hatched at the same time as the controls, were fed a similar artificial diet as the controls, but it was supplemented with 0.02% OP (w/w). The healthy controls have cuticles that are bright green with black and white diagonal stripes; the small inhibited ones are covered with dried-up, yellowish remnants of the cuticle that, in an early instar, they could not shed. Because of this, inhibited larvae are unable to resume feeding: During the molting

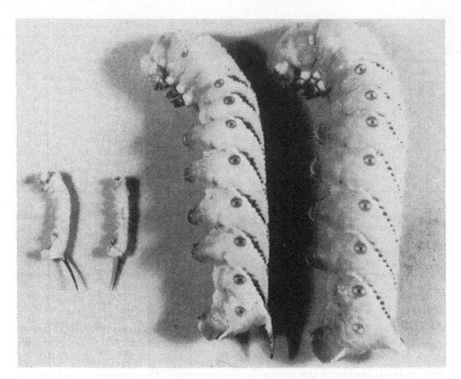

Figure 1. Molt-inhibited *Manduca* larvae. On the left are shown *Manduca sexta* larvae reared on an artificial diet containing 0.02% OP. On the right are two controls of the same age reared on regular artificial diet. The controls are of the proper size for the end of instar IV.

phase, they do not respond to stimuli inviting them to feed during the intermolt [6], and therefore, they remain the size they had achieved (seemingly the end of instar I and II) when protease was sufficiently inhibited to prevent the chitinases from attacking the chitin component of the cuticle. This hypothetical explanation is partially confirmed in that a novel and so far unidentified peak appears in the scant molting fluid from chelator-inhibited larvae when it is analyzed in the fluorimeter. With an exciting wavelength of 298 nm, no free OP was detected at 334 nm, but a new emission peak near 400 nm was seen, which is consistent with emission by a metal chelate at the exciting wavelength employed. A similar emission peak was observed when molting fluid from 2,2′-dipyridyl-inhibited *Manduca* larvae was analyzed under the same conditions in the fluorimeter. A contributing factor may be that they are presumably incapable of efficiently producing new cuticle absent breakdown products (Bade, manuscript in preparation). Figure 2 compares Growth Indices for uninhibited *Manduca* controls to various levels of 8-HQ and OP fed to *Manduca* larvae at the same age. Note that the high level of 8-HQ sulfate in later instars leads to growth indices that are slightly higher than controls. This may not be statistically significant for the *Manduca* measurements, but it was confirmed for rats.

The data shown in Figures 1 and 2 were obtained with *Manduca* larvae fed on a chelator-laced diet from the hatchling stage. In another experiment, 12 normal early instar III larvae were transferred to 0.02% OP-supplemented diet. None molted successfully into instar V. Larvae transferred similarly to 0.05% 2,2'-dipyridyl-laced diet were unsuccessful in completing the molt from instar III→IV.

We examined chelators of every type listed in the Eastman-Kodak catalog; they are given in Table 2. It was found that inhibition of the type show in Figures 1 and 2 is restricted to the nitrogen-nitrogen type of chelators represented by OP, 2,2'-dipyridyl (2,2'-Dp) and 2,2'-biquinoline (patent pending). The oxygen-nitrogen chelator 8-HQ, which is an efficient inhibitor of neutral metal chelator-sensitive pro-

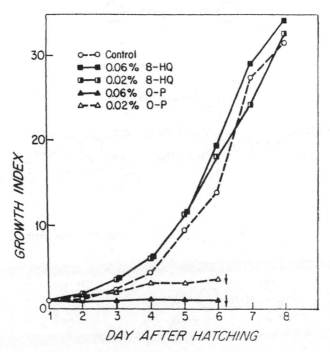

Figure 2. Growth index for *Manduca* larvae reared on various diets. Effects on growth indices (see text) are shown. ↓ terminated.

Table 2. Chelators.

Structures

NITROGEN–NITROGEN

1,10-Phenanthroline

2,2′-Dipyridyl

2,2′-Biquinoline

NITROGEN–SULFUR

Dithizone

OXYGEN–NITROGEN

EDTA

Naphthol Blue Black

Eriochrome Black T

3-Hydroxyquinoline

Table 2. (continued).

Structures

OXYGEN-OXYGEN

Quercetin

---

tease *in vitro*, and Eriochrome Black T (EBT), which has a double dose of a similar oxygen-nitrogen chelating system but is presumably too bulky to penetrate into the molting space, do not outright inhibit molting when fed to larvae, but they slow the development process. The O-N chelator Naphthol Blue Black (NBB) is inactive when fed. Note that the O-N chelator EDTA differs from the effective N-N and O-N linkages in not having aromatic rings involved in the chelation. EDTA also did not inhibit molting, but retarded *Manduca* growth somewhat.

Table 3 lists growth indices for *Manduca* larvae for 2,2′-dipyridyl and some other chelators. Note that for EDTA, arrival at the wandering stage, i.e., when the full grown larvae leave the food and seek out a place to pupate, is somewhat delayed, and growth indices are smaller, although pupation is achieved. A similar though less pronounced effect is seen with Eriochrome Black T. The EDTA effect is of interest because this was one of two chelators used in a report on "Chelating Agents Suppress Pupation of the Cabbage Looper" [7]. Comparison with the results presented in Table 3 suggests that the effect on the cabbage looper, visible in photographs of twisted pupae, was perhaps more due to the very high levels of metal chelate fed (up to 0.5%), rather than to ingestion of the chelators themselves. We compared the effect on the larval growth index of feeding two levels of OP to *Manduca* with what occurred when feeding the same levels of OP complexed with ferrous ions; this prolonged life somewhat but in the long run did not allow successful molting or pupation in *Manduca* sexta. Results are plotted in Figure 3.

Using eight paired groups of laboratory rat litter mates, we fed *ad libitum* regular rat chow (controls) and rat chow supplemented with

Table 3. Growth indices for Manduca larvae.

Day	2,2'DP (0.02%)	EDTA (0.02%)	EDTA (0.05%)	EBT (0.05%)	NBB (0.05%)	Controls
1	1.1	—	—	—	—	1.1
2	1.9	3.2	2.9	—	—	3
4	2.1	5.0	4.8	—	—	4.3
6	All moribund	—	11	12	11	14
9		26	—	—	24	24
10		—	25	36	—	34
11		41	—	—	40	38
12		—	40	58	—	43
13		85	—	—	80	71
15		—	65	115	142	141
18		Most wandering	120	All wandering	All wandering	(Most wandering)
20		All wandering	All wandering	—	—	All wandering
24		All pupated	N.d.	All pupated	N.d.	All pupated

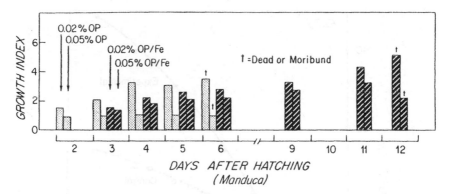

Figure 3. Effect of supplementing OP-laced diet with Fe^{2+}. Bar graph shows slight improvement of survival rate when inhibiting OP-diet is fed. See text for further explanations.

0.02% OP (experimental). Between day 8 and day 63, when the experiment was terminated, the experimental animals grew 20% faster than the controls. Figure 4 shows this effect. At the conclusion of the experiment, blood samples were obtained by cardiac puncture and iron content was compared, and the rats were grossly autopsied. Since OP is a potent iron chelator, anemia was considered a possibility. No unusual fat deposits or other abnormalities were noted in the experimental animals, and Fe^{+2} content of blood was similar in the two groups. Iron levels are shown in Table 4. In the rats, OP may instead have served like an antibiotic in a feedlot since these rats were confined in cages and were probably not as happy as rats out on their own, just as cattle confined under crowded conditions. At any rate, the chelator-fed rats ex-

Table 4. Blood iron content of rats fed OP.

	Diet	
Control (mg% Fe)		0.02% OP (mg% Fe)
17.68		19.41
19.76		19.63
18.55		18.64
17.85		lost
19.42		18.52
20.46		10.05
18.95 ± 1.1		19.25 ± 0.6

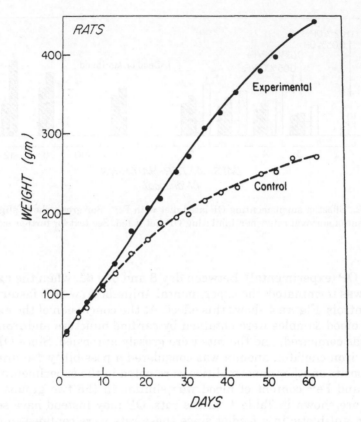

Figure 4. Weight of rats on experimental diet vs. controls on regular laboratory chow. Experimental rats were fed on OP-laced rat chow and grew faster. See text for further details.

perienced no evident harm, which is encouraging if one wishes to base an insecticide, or rather an insectistat, for chewing insects on use of chelators. This result, which may also be true for birds and other vertebrates because they share enzyme systems with rats and not insects, was expected on the basis of the unique role and identity of the enzymes involved in insect molting. Of course, more experimentation is needed before it is known with certainty which process(es) is/are affected in the molt, especially where major changes in insect form occur during the life cycle.

Effects of chelator feeding on the larval stages of some other insect orders were also examined. The results make it clear that pupation or pupariation are critical transition stages vulnerable to blocking. *Drosophila melanogaster* is a genus of fruit flies (order Diptera) in which the three larval stages live in a wet medium, which in the old days was made from mashed bananas and baby cereal. In the experiments, *Drosophila* medium was supplemented with the two levels of OP that had been found effective with *Manduca*, viz., 0.02% and 0.05%, and the cultures were set up in cotton-stoppered flasks seeded with three sexed pairs of adults each. The lower OP level permitted completion of larval instar III but visibly interfered with pupariation: The (often miniature) larvae ascended the glass wall of the flasks but were unable to complete pupariation; it looked as though they dried up on the glass. The higher level in the medium did not permit appearance of instar III. Based on these results, open bowls containing *Drosophila* laced with OP and set out for *Drosophila* adults were effective in wiping out the *Drosophila* infestation that the genetics lab wished off on our *Manduca*-growing room at least once per semester. 2,2'-Dipyridyl was as effective as OP in preventing pupariation. But in the genus *Drosophila* (order Diptera), unlike what was observed with *Manduca*, which belong to order Lepidoptera, EDTA accelerated the appearance of the next generation of (normal) adult flies. Some results are summarized in Table 5.

Feeding experiments were also tried with a *Tenebrio* sp. (order Coleoptera). Commercial rabbit chow (100 g) was slurried with 250 ml water in which chelator at 0.06% (w/w) had been dissolved. After dry-

Table 5. *Effect of chelators on* Drosophila *cultures.*

Compound	Larvae	Ascent	Pupariation	Adult New Gen.
2,2'-Dp				
0.05%	Few	None	No	None
0.03%	10% of controls	Miniatures	No	None
OP				
0.05%	Few	None	No	None
0.03%	10% of controls	Miniatures	No	None
Fe^{+2}/OP				
Chelate	Tracks seen	Yes	Yes	27% of controls
EBT				
0.05%	Yes	Yes	Yes (much size variation noted)	50% of controls
EDTA				
0.05%	Yes	Yes	Yes	Yes[a]

[a]EDTA accelerated emergence of instar III, pupariation, and emergence of adult flies. In one experiment, heavy deposition of puparia was noted, and ten adults had hatched in three EDTA cultures before emergence of the first III in any of the control bottles.

ing, the chow was coarsely crumbled and spread on the bottom of empty Crisco cans. These were filled with shredded newsprint. Half of a raw potato was wired into the cover to maintain favorable moisture conditions, and each culture was seeded with two adult beetle pairs. Cultures were set up with no inhibitor, with OP, 8-HQ, and EDTA. The inhabitants of the culture supplemented with 8-HQ were very frisky, rather like the experimental rats, and this culture was not continued. Larvae in the other cultures were periodically examined, counted, measured, and a Larval Development Index was calculated according to the formula:

$$\text{Larval Development Index} = \frac{\text{Lengths} \times \text{Widths}}{\text{Number of larvae per culture}}$$

The findings were, in brief, that presence of OP in the chow prevented pupation and emergence of a new generation of adults. EDTA, however, seemed to confer a curious sort of immortality on the larvae present: The majority continued, without further growth or pupation, in large size and considerable numbers, to the termination of the experiment. Results are summarized in Table 6. Reduction in growth indices in later stages indicates the arrival of new, small larvae, i.e., the offspring of the second generation, together with some dying off. The outcome of the experiments with *Tribolium* may have been affected by the presence of the raw potato half; nevertheless, it indicates that further experiments with selected chelators should be pursued. Expected is an environmentally benign insectistat that interferes only with the health and happiness of chewing pests, from clothes moths to gypsy moths.

Table 6. Effect of chelator ingestion on mealworms.

Month of Exper.	Larval Development Index			Pupae			Adults (2nd Gen.)	
	Control	EDTA	OP	Control	EDTA	OP	Control	EDTA
Sept. 30	44	64	36					
Oct. 10	49	71	30					
Oct. 30	75	95	31					
Nov. 15	110	128	41					
Nov. 30	125	153	50					
Dec. 10	121	162	46					
Dec. 20	129	159	49	Yes	No	No		
Jan. 15	120	126	42	Yes	No	No		
Feb. 15	67	N.d.	46	Yes	No	No	Yes	No
Mar. 15	70	128	43	Yes	No	No	Yes	No

CONCLUSIONS

1. Insects employ enzyme systems that appear to be nearly unique.
2. Protease inhibitors that work in the test tube will inhibit chitinase activity and hence successful completion of the molting process, if they are able to reach the ecdysial space that forms between old and new cuticle.
3. Molting and pupation are vulnerable steps in the insect life cycle worthy of further study. Fuller understanding of this and other enzyme systems that are narrowly restricted in distribution to certain phyla and are vital to pest metabolism is expected to lead to development of effective, specifically targeted pesticides with negligibly harmful side effects.
4. An effective molt inhibitor had a stimulating effect on rats when fed at the same level that proved inhibitory when insects were ingesting it.

ACKNOWLEDGEMENTS

This work was previously supported by National Institute of Health (NIH) and National Science Foundation (NSF), before work on environmental matters became unfashionable. Grateful acknowledgement is made to my co-workers: Alfred Stinson, Senior Research Associate; Michael Kosmo, J. J. Shoukimas, Ivan Boyer, Kerry Hickey, and L. Lapierre, students; Nehad A.-M. Moneam, postdoctoral associate, and G. R. Wyatt, Ph.D. thesis advisor.

REFERENCES

1. Smith, D. S. 1968. *Insect Cells, Their Structure and Function.* Edinburgh: Oliver and Boyd.
2. Bade, M. L. and A. Stinson. 1978. "Digestion of Cuticle Chitin during the Moult of *Manduca sexta* (Lepidoptera: Sphingidae)." *Insect Biochem.* 9:221–231.
3. Bade, M. L. and J. J. Shoukimas. 1974. "Neutral Metal Chelator-Sensitive Protease in Insect Moulting Fluid." *J. Insect Physiol.* 20:281–290.
4. Katzenellenbogen, B. S. and F. C. Kafatos. 1971. "Proteinases of Silkmoth Moulting Fluid: Physical and Catalytic Properties." *J. Insect Physiol.* 17:775–800.
5. Yamamoto, R. T. 1969. "Rearing the Tobacco Hornworm on an Artificial Diet." *J. Econ. Ent.* 62:1427–1431.
6. Chapman, R. F. 1971. *The Insects: Structure and Function.* p. 685. New York: American Elsevier.
7. Sell, D. K. and C. H. Schmidt. 1968. "Chelating Agents Suppress Pupation of the Cabbage Looper." *J. Econ. Entomol.* 61:946–949.

CHAPTER 10

Properties of Insect Chitin Synthase: Effects of Inhibitors, γ-S-GTP, and Compounds Influencing Membrane Lipids

MICHAEL LONDERSHAUSEN AND ANDREAS TURBERG

Institute for Parasitology

Bayer AG

51368 Leverkusen, Germany

MONIKA LUDWIG, BÄRBEL HIRSCH AND MARGARETHE SPINDLER-BARTH

Lehrstuhl für Hormon- und Entwicklungsphysiologie

Heinrich-Heine-University

40225 Düsseldorf, Germany

INTRODUCTION

Benzoylphenylureas interfere with the formation of chitin [1], but the underlying biochemical mechanism still remains unclear. During the last 10–15 years thousands of benzoylphenylureas have been synthesized by chemical companies [2], resulting in about five to eight products useful for application in the field of crop protection and animal health since the launch of dimilin. Extended biological screening recently led to the discovery of tick-specific acaricidal benzoylphenylureas (BPUs) [3]. This, on the one hand, shows that still new specific compounds can be identified within the enormous pool of BPUs; on the other hand, it seems now that this class is more or less exploited with regard to their resistant breaking properties and improvement of activity. This situation is also reflected by the drastic decrease in number of

155

patent applications for BPU-type compounds [2]. So far, target site-directed investigations on animal health parasites gave no clear structure-activity relationship, but in combination with investigations on penetration and metabolism of benzoylphenylureas, it seems likely that pharmacokinetic properties and metabolic stability are more important for biological activity than binding to a presumed target site [4–6]. Since chitin formation is still a very interesting target for parasite growth regulators, a cell line from *Chironomus tentans* was used as a model system to investigate arthropod enzymes directly without interference by penetration barriers, hormonal regulation, or problems of synchronization of the test animals [7]. In this context, the insect cell line will be an interesting tool to investigate the primary mode of action of various inhibitors, including the benzoylphenylureas. Used in biochemical screening systems, it will facilitate the molecular identification of new targets for alternative arthropod-specific growth regulators.

BIOLOGICAL ACTIVITY OF CHITIN SYNTHESIS INHIBITORS

For evaluation of structure-activity relationships a series of different compounds was investigated displaying selective activities against animal health parasites and arthropods relevant as hygiene pests and vectors. In previous investigations it was shown, that 2,6-difluoro- and 2-chloro-benzoyl-analogues with 4-chloro-, 4-fluoro-, or 4-trifluorome-thoxy-substituents belonged to the most potent compounds [8]. Further analyses therefore concentrated on a few selected BPUs varying at the phenyl- and the pyridoxyloxyphenyl moiety, respectively. For these compounds species-specific differences were observed (Table 1). Although diflubenzuron, triflumuron, and SIR14591 displayed pronounced growth-retarding effects against all species tested, lufenuron revealed an even higher activity against the arthropod species but was inactive against the nematode larvae. SIR32046 and teflubenzuron, both carrying similar substituents at the phenyl residue, were different in activity by a factor of 10–100 against *Aedes aegypti* and *Lucilia cuprina,* were more or less comparably efficient against flea but inactive against nematode larvae. For flufenoxuron and chlorfluazuron, both oxyphenyl derivatives, similar results were obtained. Chlorfluazuron was more active against *Aedes aegypti* and *Lucilia cuprina* but less active against *Ctenocephalides felis* than flufenoxuron. The meta-substituted pyridoxyloxyphenyl compound fluazuron was completely inactive against insects but exhibited highly selective effects against the cattle tick *Boophilus microplus* [3], indicating that the structure-activity relationship for acaricides is significantly different from insec-

Table 1. Activity of different BPUs on veterinary parasites.

Compound	R₁	R₂	R₃	R₄	LD_{90}			
	R_1	R_2	R_3		Aedes aegypti Larvae and Pupae[a] ($\mu g\ ml^{-1}$)	Lucilia cuprina Larvae and Pupae[b] ($\mu g\ ml^{-1}$)	Ctenocephalides felis Larvae and Pupae[c] ($\mu g\ g^{-1}$)	Haemonchus contortus Larvae[d] ($\mu g\ ml^{-1}$)
Diflubenzuron	H	F	F		~0.1	100	100	~100
Triflumuron	H	Cl	H		0.01–0.1	100	100	~100
SIR32046	F	Cl	H		0.001–0.01	10–100	~100	>100
Teflubenzuron	H	F	F		0.01–0.1	1,000	100	>100
SIR14591	H	F	F		0.01–0.1	10	~100	~100

(continued)

157

Table 1. (continued).

Compound	R₁	R₂	R₃	R₄	LD₉₀			
					Aedes aegypti Larvae and Pupae[a] (μg ml^{-1})	Lucilia cuprina Larvae and Pupae[b] (μg ml^{-1})	Ctenocephalides felis Larvae and Pupae[c] (μg g^{-1})	Haemonchus contortus Larvae[d] (μg ml^{-1})
Lufenuron	H	F	F		0.01–0.1	10	1–10	≥100
Flufenoxuron	H	F	F		0.1–1	100–1,000	10	100
Chlorfluazuron	H	F	F		0.001–0.01	10–100	~100	>100
Fluazuron	H	F	F		≥10	>1,000	>1,000	n.d.

[a]All developmental stages were tested in 50 ml water (21°C–25°C) using 0.0001 to 10 μg/ml of the compound under investigation. Three replicates were done for each concentration, and activity was evaluated at days 1, 3, 7, 14, and 21. Animals were fed daily with Tetramin®.
[b]All developmental stages were tested using minced horse meat (20 g/assay) treated with 2 ml of test compound containing 0.1 to 1000 μg/ml. Twenty to fifty first larvae were placed on meat and incubated at 26°C and 75% relative humidity untill pupation.
[c]Flea eggs were placed on a dried cattle blood (supplied with yeast extract and sea sand) meal sample containing 10 to 1,000 μg/g of the compound under investigation. Incubation was carried out at 25°C and 85% relative humidity and mortality of flea larvae and pupae was determined.
[d]Larvae were obtained essentially as described by Roberts and O'Sullivan [19]. Compounds were added to the medium to yield final concentrations of 1 to 100 μg/ml.

158

ticides. Toxicological studies showed differences of diflubenzuron and chlorfluazuron in *Heliothis virescens* and *Spodoptera littoralis*. Chlorfluazuron revealed an at least tenfold longer half-life in larvae and less excretion of insecticide metabolites during the treatment period [4]. This means that metabolism, which might differ substantially between species, is one controlling factor for insecticidal activity [5]. Tests under conditions where oxidative metabolism of BPUs was suppressed by piperonylbutoxide in *Chilo suppressalis* indicated that the interaction with a receptor site may play a key role for biological activity [9]. The fact that BPUs with a wide variety of mono- and multi-substituted anilide moieties are represented by a single quantitative correlation equation supports this hypothesis [9]. To evaluate structural parameters that might influence the biological activity of BPUs, the in vitro metabolism of two [^{14}C]-labeled derivatives (triflumuron and SIR14591) were investigated for two insect species and an insect cell line. As shown in Table 2 the biologically more active SIR14591 was significantly less metabolized in *Locusta migratoria* and *Lucilia cuprina* epithelial cell homogenates than triflumuron. The less pronounced metabolism of SIR14591 in correlation with its better larvicidal activity seems to indicate that for the above-mentioned species, degradation and excretion of BPUs also play a key role for primary biological activity as was clearly demonstrated for *Spodoptera littoralis* after treatment with flufenoxuron, teflubenzuron, and diflubenzuron [6]. No degradation could be detected in the insect cell line (Table 2).

Table 2. Metabolic degradation of triflumuron and SIR14591.

Organism	% Metabolism[a] of [^{14}C] Radiolabelled	
	Triflumuron	SIR14591
Locusta migratoria	5.7	1.3
Lucilia cuprina	23.1	9.9
Chironomus tentans (cell culture)	<1	<1

[a]Metabolism of triflumuron and SIR14591 was determined after homogenization of 100 mg insect tissue or cells in 2 ml Grace's medium at 37°C. Penicillin G, gentamycinsulphate, and streptomycinsulphate (25 µg/ml each) were added and the reaction stopped after 72 h by extraction with 7 × 1 ml ethylacetate. The samples were evaporated to a volume of 500 µl and subjected to TLC (Silicagel 60, 20 × 20) with solvent system benzole/diphenylether (5/1). The decrease of "mothercompound" was measured using an Isomess DC Scanner IM 3000.

This allows the investigation of chitin synthesis inhibitors without metabolism or interference with pharmacokinetic influences.

EFFECTS OF INHIBITORS, γ-S-GTP, AND COMPOUNDS INTERACTING WITH MEMBRANE LIPIDS ON CHITIN BIOSYNTHESIS IN AN INSECT CELL LINE FROM *CHIRONOMUS TENTANS*

The lipid environment is essential for physiological function of chitin synthase in fungi [10] and yeast [11], where not only the stability [12] but also enzymatic parameters such as allosteric activation by GlcNAc were reduced upon detergent treatment [13]. In general, solubilization attempts with different detergents deteriorate insect chitin synthase activity with only low concentrations of octyl-β-D-thioglucopyranoside (OTG) being compatible (Figure 1). Although similar effects were observed with chitin synthase from fungi, especially with higher detergent concentrations, solubilization of the enzyme with preserved activity is possible [13], whereas chitin synthase from *Chironomus* cells was far more sensitive to detergent treatment (Figure 1). The importance of phospholipids for insect chitin synthesis could also be confirmed for *Chironomus* cells (Figure 2). After OTG treatment the decrease in activity could be restored at best with phosphatidylcholine-1 and phosphatidylserine, whereas other phospholipids were inefficient, in-

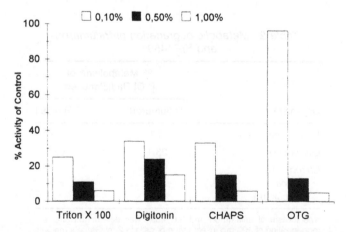

Figure 1. Effects of detergents on chitin synthesis in *Chironomus* cell homogenates. Chitin synthesis was measured using 70 nM [³H] UDP-GlcNAc as substrate [14] after treatment with detergents for 12 h at 30°C. Control value corresponds to 0.84 pmol/mg protein/12 h in homogenates. CHAPS (3-[(3-cholamidopropyl)-dimethylammonio]-1-propanesulfonate); OTG (octyl-β-D-thioglucopyranoside); Triton ×100 (octylphenol-poly-ethyleneglycolether, *n* ~ 10).

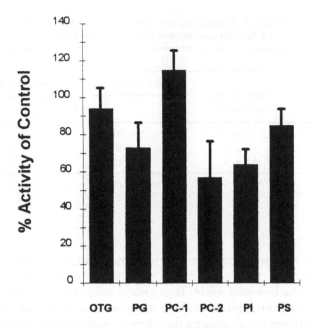

Figure 2. Influence of different phospholipids on chitin synthesis after detergent treatment. Chitin synthesis was measured using 70 nM [³H] UDP-GlcNAc as substrate and 0.2 mg/ml of indicated phospholipids. Prior to incubation, the homogenates were treated with 0.1% OTG for 24 h at 30°C. Control activity corresponds to 0.12 pmol/mg protein/ 12 h. PC-1 (phosphatidylcholine-1 = β-acetyl-γ-hexadecyl-phosphatidylcholine); PC-2 (phosphatidylcholine-2 = distearyl-phosphatidylcholine); PG (phosphatidylglycerol); PI (phosphatidylinositol); PS (phosphatidylserine).

dicating a selective requirement of chitin synthase for these membrane components. In addition, the treatment with phospholipases (Table 3) clearly decreased membrane bound enzymatic activity without significant solubilization. The importance of lipid environment and the deleterious effects of detergents have impaired the further characterization and the isolation of the enzyme so far.

After differential centrifugation (Table 4), chitin synthase was more or less equally distributed between the 300-g, 16,000-g and 100,000-g pellets, whereas in the 100,000-g supernatant only small amounts of substrate were incorporated, indicating particle bound biosynthesis of chitin. A comparable distribution of chitin synthase was observed in other cell-free preparations of arthropods [16,17]. Since the substrates GlcN and UDP-GlcNAc were accepted by all particle bound chitin synthases from *Chironomus tentans* cells, a common pathway for both precursors is assumed [15,18].

The distribution of chitin synthase activity in all membrane frac-

Table 3. Effect of phospholipase treatment on chitin synthesis in Chironomus tentans cell homogenates.

Treatment[a]	Total Activity % Control	Chitin Synthesis	
		16,000 g Pellet	16,000 g Supernatant
		% Total Activity	
None	100 ± 10	73 ± 12	27 ± 5
80 mU PL-C	48 ± 6	43 ± 8	57 ± 8
80 mU PL-D	43 ± 6	58 ± 6	42 ± 6

[a]Prior to incubation with 70 nM [³H] UDP-GlcNAc the homogenates were treated with phospholipases (80 mU) for 24 h at 25°C and then subjected to differential centrifugation. Control activity corresponds to 0.12 pmol/mg protein/12 h. PL-C (phosphatidyl-D-myo-inositol inositol phosphorylase); PL-D (phosphatidylcholine-phosphatidohydrolase).

tions tested may indicate that chitin formation takes place at various compartments in the cell. This is supported by data from *Artemia salina,* where a two-step model is proposed with initiation of the carbohydrate chain taking place at the endoplasmatic reticulum and transport of precursors to the plasma membrane, where final polymerization to high molecular weight chitin occurs [20,21].

So far, inhibition of precursor transport [22], proteolytic activation of a chitin-synthase zymogen [23], direct inhibition of chitin synthase [24], indirect hormonal effects [25], and effects on cell membranes in combination with vesicle transport [26] have been discussed as the primary mode of action of BPUs. The investigation of chitin synthase of

Table 4. Distribution pattern of chitin synthesis after differential centrifugation.

Fraction	Incorporation into Chitin	
	[³H] UDP-GlcNAc	[³H] GlcN
	pmol/mg protein/12 h (% of total activity)	
300 g Pellet	1.75 ± 0.21 (35 ± 6)	3.25 ± 0.36 (33 ± 5)
16,000 g Pellet	1.60 ± 0.29 (37 ± 4)	2.75 ± 0.51 (38 ± 8)
100,000 g Pellet	1.61 ± 0.29 (14 ± 3)	2.50 ± 0.43 (15 ± 4)
100,000 g Supernatant	0.75 ± 0.18 (15 ± 3)	1.00 ± 0.29 (14 ± 3)

Homogenates were subjected to differential centrifugation. The fractions were incubated with 70 nM [³H] UDP-GlcNAc or [³H] GlcN for 12 h at 30°C, and chitin production was determined according to Ludwig et al. [15].

Chironomus tentans cells indicate that the enzyme generally displays properties similar to other arthropod chitin synthases [27]. In combination with the above-mentioned advantages, investigations with isolated epithelial tissue offer the opportunity to deduct the primary effects of BPUs directly from their interaction with the target cell. The underlying mechanisms are poorly understood compared with yeasts and fungi. Investigations using tissue preparations, imaginal discs, organ cultures, or whole animal bioassays are complicated by time-consuming preparations, metabolic degradation, as well as problems of synchronization of the physiological stage of the experimental animals as outlined earlier [7]. This is of particular importance since the molting cycle markedly influences chitin synthesis [28].

To evaluate whether the insect cell line from *Chironomus tentans* exhibits the inhibition pattern typical for arthropod chitin synthesis, various inhibitors were applied (Table 5, Figure 3). The antifungal agent nystatin revealed only nonspecific effects at high concentrations. The compound inhibits chitin synthesis in fungi in a noncompetitive manner, presumably caused by nonspecific membrane sterol binding and forming of ion channels [29,30]. The macrotetrolide ionophore dinactin [31] and the protein synthesis inhibitor borrelidin [32] also inhibited only at high concentrations, indicating that this insect cell line is indeed a sensitive indicator for specific effects on chitin synthesis.

As expected, polyoxin D, a competitive inhibitor of fungal and arthropod chitin formation [33], impairs chitin synthesis also in the insect cell line with high sensitivity (Table 5, Figure 3). Tunicamycin, which was described as a noncompetitive inhibitor of uridine-5′-diphospho-*N*-acetyl-β-D-glucosamine (UDP-GlcNAc):dolichyl-*P*-GlcNAc-1-P transferase [34] preventing the formation of the first *N*-acetyl-β-D-glucosamine (GlcNAc) lipid unit, but not the addition of a second GlcNAc or

Table 5. Effect of inhibitory compounds on chitin synthesis in homogenates of Chironomus tentans cells.

Compound	IC_{50} (mM)[a]
Borrelidin	0.9
Dinactin	0.3
Nystatin	1.3
Polyoxin D	0.0001
Tunicamycin[b]	0.01

[a]Chitin synthesis was determined [14] using [^{14}C] UDP-GlcNAc as substrate; incubation conditions were according to Ludwig et al. [15].
[b]Mixture of Tunicamycin: A (5%), B (34%), C (40%), and D (18%).

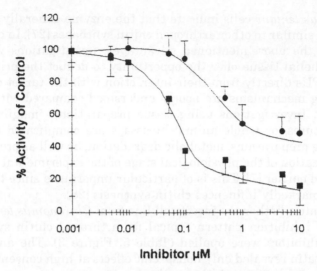

Figure 3. Inhibition of chitin synthesis in homogenates of *Chironomus tentans*. Control value corresponds to 0.21 pmol/mg protein/12 h. Polyoxin D (■), tunicamycin (●).

a mannose residue, also reduced chitin formation in the cell line. In homogenates from *Chironomus* cells, reduction by tunicamycin never exceeded 60–70%, whereas inhibition of chitin formation in intact cells was nearly complete [35]. During shorter incubation times as used for homogenates, the start of new carbohydrate chains, which is tunicamycin sensitive, seems to be less important compared with chain elongation. In addition, there might exist initiation reactions independent of amino-sugar-lipid intermediates. In contrast to the high molecular weight chains of chitin of arthropods, so far, for fungal chitin synthases only a one-step type of chitin synthesis with relatively short chains of similar length was described [36].

As already known from insect tissues, BPUs inhibit chitin formation only in intact cells [37]. In contrast to the biological activity under in vivo conditions, all BPUs tested exhibit more or less the same inhibitory potential, with the exception of teflubenzuron (Table 6). These results indicate that significant discrepancies in biological activity against target insect species might reflect distinct metabolic capacities and pharmacokinetic properties [6] rather than binding to a chitin-synthase inhibition site. Similar conclusions were drawn for chlorfluazuron, teflubenzuron, and flufenoxuron when compared in topical application and injection bioassays against *Spodoptera* species under conditions where oxidative metabolism was at least partially sup-

Table 6. *Effect of different BPUs on chitin synthesis of* Chironomus *using intact cells and cell homogenates.*

Compound (1 μM)	Chironomus tentans % Inhibition of Chitin Synthesis	
	Intact Cells[a]	Cell Homogenate[b]
Diflubenzuron	38 ± 11	3 ± 13
Triflumuron	67 ± 7	−2 ± 8
Teflubenzuron	−5 ± 2	−5 ± 16
SIR14591	40 ± 14	−11 ± 4
Chlorfluazuron	32 ± 5	13 ± 2

[a]Chitin synthesis was measured using [³H] GlcN as substrate according to Londershausen et al. [14].
[b]Chitin synthesis was measured using [¹⁴C] UDP-GlcNAc as substrate according to Ludwig et al. [15].
Control activity corresponds to 2.5 pmol/mg protein/12 h for intact cells and to 1.6 pmol/mg protein/12 h for cell homogenate.

pressed by piperonylbutoxide [4,6]. In addition, the disruption of cell integrity eliminates the effects of BPUs on chitin synthesis in *Chironomus* cells also. Since chitin synthesis still occurs in homogenates, chitin synthase itself is not the target of these inhibitors (Table 6).

As shown in Table 7, γ-S-GTP, an analogue of GTP, which is not metabolized, stimulates chitin synthesis in *Chironomus* cells. Moreover, the inhibitory action of triflumuron was counteracted by γ-S-GTP, depending on the dose applied. Ultrastructural investigations of *Spodop-*

Table 7. *Effect of* γ-S-GTP *and triflumuron on chitin synthesis in* Chironomus tentans *cells.*

Compound	Concentration (μM)	Chitin Synthesis[a] (% of Control)
γ-S-GTP	10	72 ± 13
	100	152 ± 34
	200	217 ± 31
Triflumuron	1	35 ± 7
γ-S-GTP + Triflumuron	10 + 1	112 ± 13
	100 + 1	320 ± 87

[a]Chitin synthesis was measured according to Londershausen et al. [14]. Compounds under investigation were added and cells were kept for 4 days until harvesting and sample preparation. Control activity corresponds to 3.2 pmol/mg protein/24 h.

tera [38] and *Lucilia* [39] cuticle demonstrate the accumulation of dense deposits and electron-lucent globular bodies in the subcuticle after treatment with diflubenzuron or flufenoxuron. It is tempting to speculate that BPUs are involved in G-protein-dependent intra- and/or extracellular vesicular transport of precursors. Regarding intracellular vesicles, this is in accordance with the effects of diflubenzuron on vesicular calcium uptake [26] and cAMP-dependent protein kinase [40] and may further support the two-step mechanism of chitin synthesis in arthropods as proposed for *Artemia salina* [21].

SUMMARY

The biological activity of the most important benzoylphenylurea derivatives against animal health parasites was determined for different developmental stages of the yellow fever mosquito *Aedes aegypti*, the sheep blowfly *Lucilia cuprina*, the cat flea *Ctenocephalides felis*, and the sheep intestinal nematode *Haemonchus contortus*. No clear structure-activity relationship for antiparasitic efficacy could be determined for the tested BPUs, but indications were found demonstrating the importance of metabolism and pharmacokinetic properties for biological activity. The investigation of the cell line from *Chironomus tentans*, which was shown to be devoid of any inhibitory metabolism, revealed no differences in inhibition potency at the target site. But it could be demonstrated that the inhibitory effects were essentially dependent on intact cell structure. Different inhibitors, primarily known from their effects on the fungal enzyme as well as compounds affecting exocytotic processes and membrane lipid environment, were therefore investigated using the insect cell line. Measurement of the dose-related inhibitory effects of tunicamycin and polyoxin D indicated chitin synthases with different properties; one might be responsible for chitin chain initiation, and another for chain elongation. GTP binding proteins are known to mediate exocytosis, which is assumed to play a role in secretion of chitin during cuticle synthesis. At 100 μM and above the investigation of the unmetabolizable GTP analogue, γ-S-GTP, revealed a stimulatory effect on chitin synthesis. When tested in combination with γ-S-GTP, triflumuron acted synergistically with 100 μM γ-S-GTP and restored its depressive activity at 10 μM. This indicates that exocytosis of chitin intermediates is of relevance not only for cuticle synthesis in arthropods but is related to the mode of action of benzoylphenylureas also. To localize and characterize chitin synthase within its membrane environment, different detergents and phospholipids were investigated. It could be shown that chitin synthase was primarily bound to plasma and microsomal membranes. The enzyme of *Chirono-*

mus tentans in general was much more sensitive towards treatment with detergents than fungal enzymes, with OTG being the most compatible. The decrease after treatment with OTG could at least partially be restored by addition of phosphatidylcholine or phosphatidylserine, whereas other phospholipids were inefficient. The treatment with phospholipases resulted in a significant loss of enzymatic activity, and no solubilization of chitin synthase was achieved.

CONCLUSION

So far, two steps in arthropod chitin synthesis have been identified, one sensitive to tunicamycin and another to benzoylphenylureas. This clearly discriminates chitin formation in animals from that in yeasts and fungi and demonstrates that different targets for interference besides chitin synthase itself can be used successfully. A more detailed understanding of the underlying biochemical processes involved in arthropod chitin synthesis may lead to additional targets suited for pesticide screening of alternative growth regulators. Simple model systems such as insect cell lines may facilitate these investigations. Compared with fungal and yeast chitin synthases, where different structural genes have been cloned and a detailed knowledge about processes involved in chitin formation exist, only little is known in arthropods and nematodes. Since classical approaches of protein purification were not successful because of the high sensitivity against detergents, screening of arthropod cDNA libraries with consensus sequences from yeast or fungi may be more promising. Since at present no inhibitor of arthropod chitin synthase itself is commercially used, although interruption with chitin formation in general is quite successful, this target deserves more consideration in the future.

ABBREVIATIONS

BPU (benzoylphenylurea); GlcN (β-D-glucosamine); GlcNAc (N-acetyl-β-D-glucosamine); IC_{50} (inhibitory concentration to suppress 50% of the initial activity); LD_{90} (concentration to kill 90% of the parasite population); UDP-GlcNAc (uridine-5'-diphospho-N-acetyl-β-D-glucosamine).

ACKNOWLEDGEMENTS

We thank Iris Schröder for skillful technical assistance and preparation of biological samples as well as Heidemarie Heim-Londershausen and Claudia Hüter for carefully writing the manuscript.

REFERENCES

1. Van Daalen, J. J., Meltzer, J., Mulder, R. and Wellinga, K. 1972. "A Selective Insecticide with a Novel Mode of Action," *Naturwiss.*, 59:312–313.

2. Naumann, K. 1994. "Neue Insektizide," *Nachr. Chem. Tech. Lab.*, 42(3):255–261.

3. Kemp, D. H. 1990. "Mode of Action of CGA 157419 on the Cattle Tick *Boophilus microplus*," *Bull. Soc. Franc. Parasitol.*, 8:1048.

4. Neumann, R. and Guyer, W. 1987. "Biochemical and Toxicological Differences in the Modes of Action of the Benzoylureas," *Pestic. Sci.*, 20:147–156.

5. Haga, T., Toki, T., Koyanagi, T. and Nishiyama, R. 1987. "Structure–Activity Relationships of Benzoylphenyl Ureas," in *Chitin and Benzoylphenyl Ureas*, eds. Wright, J. E. and Retnakaran, A., Dr. W. Junks Publishers, Chapter 6, pp. 111–129.

6. Clarke, B. S. and Jewess, P. J. 1990. "The Inhibition of Chitin Synthesis in *Spodoptera littoralis* Larvae by Flufenoxuron, Teflubenzuron and Diflubenzuron," *Pestic. Sci.*, 28:377–388.

7. Dinan, L., Spindler-Barth, M. and Spindler, K.-D. 1990. "Insect Cell Lines as Tools for Studying Ecdysteroid Action," *Invert. Reprod. Dev.*, 18:43–53.

8. Hajjar, N. P. and Casida, J. E. 1979. "Structure–Activity Relationships of Benzoylphenyl Ureas as Toxicants and Chitin Synthesis Inhibitors in *Oncopeltus fasciatus*," *Pestic. Biochem. Physiol.*, 11:33–45.

9. Nakagawa, Y., Izumi, K., Oikawa, N., Kurozumi, A., Iwamura, H. and Fugita, T. 1991. "Quantitative Structure–Activity Relationship of Benzoylphenylurea Larvicides," Part VII, *Pestic. Biochem. Physiol.*, 40:12–26.

10. Montgomery, G. W. G., Adams, D. J. and Gooday, G. W. 1984. "Studies on the Purification of Chitin Synthase from *Coprinus cinereus*," *J. Gen. Microbiol.*, 130:291–297.

11. Duran, A. and Cabib, E. 1978. "Solubilization and Partial Purification of Yeast Chitin Synthetase—Conformation of the Zymogenic Nature of the Enzyme," *J. Biol. Chem.*, 253:4419–4425.

12. Giménez, G., Gozalbo, D. and Martinez, J. P. 1991. "Stability of Chitin Synthase in Cell-Free Preparations of a Wild-Type Strain and a Slime Variant of *Neurospora crassa*," *FEMS Microbiol. Lett.*, 83:173–178.

13. Kang, M. S., Elango, N., Mattia, E., Au-Young, J., Robbins, P. W. and Cabib, E. 1984. "Isolation of Chitin Synthetase from *Saccharomyces cerevisiae*—Purification of an Enzyme by Entrapment in the Reaction Product," *J. Biol. Chem.*, 259:14966–14972.

14. Londershausen, M., Kamman, V., Spindler-Barth, M., Spindler, K. D. and Thomas, H. 1988. "Chitin Synthesis in Insect Cell Lines," *Insect Biochem.*, 18:631–636.

15. Ludwig, M., Spindler-Barth, M. and Spindler, K. D. 1991. "Properties of

Chitin Synthase in Homogenates from *Chironomus* Cells," *Arch. Insect Biochem. Physiol.*, 18:251–263.

16. Cohen, E. and Casida, J. E. 1980. "Properties of Tribolium Gut Chitin Synthase," *Pestic. Biochem. Physiol.*, 13:121–128.

17. Ward, G. B., Beydon, P. P., La Font, R. and Mayer, R. T. 1990. "Metabolism of Ecdysteroids by Chitin-Synthesizing Insect Cell line," *Arch. Insect Biochem. Physiol.*, 15:137–148.

18. Crosscurt, A. C. and Jongsma, B. 1987. "Mode of Action and Insecticidal Properties of Diflubenzuron," in *Chitin and Benzoylphenyl Ureas*, eds. Wright, J. E. and Retnakaran, A., Dr. W. Junk Publishers, Chapter 4, pp. 75–99.

19. Roberts, F. H. S. and O'Sullivan, P. J. 1950. "Methods for Egg Counts and Larval Cultures for Strongyles Infesting the Gastro-Intestinal Tract of Cattle," *Agric. Res.*, 1(1):99–102.

20. Horst, M. N. 1989a. "Molecular and Cellular Aspects of Chitin Synthesis in Larval Artemia," in *Cell and Molecular Biology of Artemia Development*, eds. Warmer, A. H., McRae, T. H. and Bagshaw, J. C., Plenum Press, New York, pp. 59–76.

21. Horst, M. N. 1983. "The Biosynthesis of Crustacean Chitin. Isolation and Characterization of Polyphenol-Linked Intermediates from Brine Shrimp Microsomes," *Arch. Biochem. Biophys.*, 223(1):254–263.

22. Mitsui, T., Nobusawa, C. and Fukami, J. 1984. "Mode of Inhibition of Chitin Synthesis by Diflubenzuron in the Cabbage Armyworm, *Mamestra brassicae* L.," *J. Pestic. Sci.*, 9:19–26.

23. Marks, E. P., Leighton, T. and Leighton, F. 1982. "Modes of Action of Chitin Synthesis Inhibitors," in *Insecticide Mode of Action*, ed. Coats, J. R., Academic Press Inc., New York, pp. 281–313.

24. Van Eck, W. H. 1979. "Mode of Action of Two Benzoylphenyl Ureas as Inhibitors of Chitin Synthesis in Insects," *Insect Biochem.*, 9:295–300.

25. Soltani, N., Delbeque, J. P., Delachambre, J. and Mauchamp, B. 1984. "Inhibition of Ecdysteroid Increase by Diflubenzuron in *Tenebrio molitor* Pupae and Compensation of Diflubenzuron Effect on Cuticle Secretion by 20-Hydroxyecdysone," *Int. J. Inv. Reprod. Dev.*, 7:323–332.

26. Nakagawa, Y. and Matsumura, F. 1994. "Diflubenzuron Affects Gamma-thio-GTP Stimulated Ca^{2+} Transport in vitro in Intracellular Vesicles from Integument of Newly Molted American Cockroach, *Periplaneta americana* L.," *Insect Biochem. Mol. Biol.*, 24(10):1009–1015.

27. Kramer, K. J. and Koga, D. 1986. "Insect Chitin: Physical State, Synthesis, Degradation and Metabolic Regulation," *Insect Biochem.*, 16(6):851–877.

28. Spindler, K. D. 1983. "Chitin: Its Synthesis and Degradation in Arthropods," in *The Larval Serum Proteins of Insects. Function, Biosynthesis, Genetics*, ed. Scheller, K., Georg Thieme Verlag Stuttgart, New York, pp. 135–149.

29. Hector, R. F. 1993. "Compounds Active against Cell Walls of Medically Important Fungi," *Clin. Microbiol. Rev.*, 6(1):1–21.

30. Gräfe, U. 1992. *Biochemie der Antibiotika,* Spektrum Akademischer Verlag, Heidelberg, Berlin, New York, Chapter 5.5.4, pp. 279–280.

31. Gräfe, U. 1992. *Biochemie der Antibiotika,* Spektrum Akademischer Verlag, Heidelberg, Berlin, New York, Chapter 4.2.2, pp. 123–124.

32. Gräfe, U. 1992. *Biochemie der Antibiotika,* Spektrum Akademischer Verlag, Heidelberg, Berlin, New York, Chapter 5.4.2.1, pp. 226–227.

33. Hajjar, N. P. 1985. "Chitin Synthesis Inhibitors as Insecticides," in *Insecticides, Progress in Pesticide Biochemistry and Toxicology,* eds, Hutson, D. H. and Roberts, T. R., John Wiley and Sons Ltd., New York, Chapter 7, pp. 275–310.

34. Heifetz, A., Keenan, R. W. and Elbein, A. D. 1979. "Mechanism of Action of Tunicamycin on the UDP-GlcNAc: Dolichyl-Phosphate GlcNAc-1-Phosphate Transferase," *Biochem.,* 18(11):2186–2192.

35. Spindler-Barth, M., Spindler, K. D., Londershausen, M. and Thomas, H. 1989. "Inhibition of Chitin Synthesis in an Insect Cell-Line," *Pestic. Sci.,* 25:115–121.

36. Cabib, E. 1987. "The Synthesis and Degradation of Chitin," *Adv. Enzym.,* 59:59–101.

37. Mayer, R. T., Chen, A. C. and DeLoach, J. R. 1981. "Chitin Synthesis Inhibiting Insect Growth Regulators Do Not Inhibit Chitin Synthase," *Experienta,* 37:337–338.

38. Lee, S., Clark, B. S., Jenner, D. W. and Williamson, F. A. 1990. "Cytochemical Demonstration of the Acylureas Flufenoxuron and Diflubenzuron on the Incorporation of Chitin into Insect Cuticle," *Pestic. Sci.,* 28:367–375.

39. Binnington, K. C. 1985. "Ultrastructural Changes in the Cuticle of the Sheep Blowfly, *Lucila,* Induced by Certain Insecticides and Biological Inhibitors," *Tissue and Cell,* 17(1):131–140.

40. Ishii, S. and Matsumura, F. 1992. "Diflubenzuron-Induced Changes in Activities of the cAMP-Dependent Protein Kinase in the Newly Molted Integument of the American Cockroach in situ and in Cell Free Conditions," *Insect Biochem. Mol. Biol.,* 22(1):69–72.

CHAPTER 11

New Applications of Chitin and Its Derivatives in Plant Protection

HENRYK STRUSZCZYK
Institute of Chemical Fibres
M. Sklodowskiej–Curie 19
90-570 Lodz, Poland

HENRYK POSPIESZNY
Institute of Plant Protection
Miczurina 20
60-318 Poznan, Poland

INTRODUCTION

Chitin is the second most abundant polymer occurring in nature. Chitosan is produced by deacetylation of chitin. These polymers are somewhat similar to cellulose in being polysaccharides, but the repeating unit of the polymer backbone is not glucose but glucosamine, the amino group of which is largely acetylated in the case of chitin and largely unacetylated in the case of chitosan. Chitin and chitosan are widely used in nature as structural materials but they do not occur in higher plant tissues [1,2]. However, they induce a broad spectrum of defensive plant responses [3]. Chitosan, a cationic polysaccharide, exhibits antifungal [4,5] and antiviral properties [6–9]. At the same time these polyaminosaccharides express special plant biostimulation behavior [10]. In this chapter, progress on the bioactivity of chitin and its derivatives, such as soluble chitin oligomers, chitosan, and chitosan derivatives, like sulfonated and carboxymethylated chitosan, in plant protection and growth stimulation is presented.

171

ANTIBACTERIAL ACTIVITY OF CHITIN AND ITS DERIVATIVES

The bacteria used in this study were *Clavibacter michiganense* subsp. michiganense, *C. michiganense* subsp. insidiosum, *Xanthomonas campestris* pv. pelargonii, *X. campestris* pv. phaseoli, *Pseudomonas syringae* pv. phaseolicola, *P. syringae* pv. tomato, *Erwinia amylovora, E. carotovora* subsp. carotovora, *Agrobacterium tumefaciens,* and *Escherichia coli.*

Bacterial growth susceptibility was determined by the minimum inhibitory concentration (MIC) method. Drops of chitin derivatives of different concentrations were applied to the surface of agarose plates containing cultures of bacteria in nutrient dextrose medium or LB medium for phytopathogenic bacteria and *E. coli,* respectively. MIC was defined as the lowest concentration of chitin derivatives that inhibited bacterial growth after overnight incubation of the agarose plates at 37°C.

In another experiment, the effect of chitin derivatives on *Pseudomonas syringae* pv. phaseolicola was tested using the hypersensitive reaction (HR) of tobacco. Mixtures of bacterium and chitin derivatives at a final concentration of 5×10^7 CFU/ml and 0.05 wt%, respectively, was injected into leaves of tobacco Xanthi nc. Suspension of the bacterium in distilled water or solutions of chitin derivatives in distilled water was used as controls.

Water-soluble chitin oligomers, chitosan, chitosan sulfate, and carboxymethyl chitosan were used in this research. Chitosan was dissolved in the acetic acid and other chitin derivatives in distilled water. The reaction of all solutions was adjusted to pH = 5.5–6.0 with potassium hydroxide. As shown in Table 1, cationic chitin derivatives, i.e., chitin oligomers and chitosan, inhibited growth of the gram-positive bacteria: *C. michiganense* subsp. michiganense and *C. michiganense* subsp. insidiosum, and gram-negative bacteria: *X. campestris* pv. pelargonii, *X. campestris* pv. phaseoli, *P. syringae* pv. phaseolicola, *P. syringae* pv. tomato, *E. amylovora, E. carotovora* subsp. carotovora, and *A. tumefaciens* at concentration of 0.01–0.3 wt%. However, both derivatives were less effective against *E. coli.* Anionic chitin derivatives, i.e., chitosan sulfate and carboxymethyl chitosan at a concentration of 1.5 wt% were not effective against any of the bacteria tested. When cationic derivatives were added to the bacteria suspension, the flocculation of them was observed. The hypersensitive reaction (HR) of plants is widely used for quick demonstration of bacterial pathogenicity [11]. When the tobbacco leaves were injected by a mixture of *Pseudomonas syringae* pv. phaseolicola and chitin derivatives, HR was prevented (Table 2).

Table 1. Inhibition of bacterial growth by chitin derivatives.

Chitin Derivatives[a]	MIC (%)[b]									
	Cmm[c]	Cmi[c]	Xcp[c]	Xcph[c]	Psph[c]	Pst[c]	Ea[c]	Ecc[c]	At[c]	Ec[c]
Chitosan	0.1	0.1	0.25	0.1	0.25	0.1	0.25	0.25	0.25	(0.5)
S-Chitosan	(1.5)[d]	(1.5)	1.5	(1.5)	(1.5)	(1.5)	(1.5)	(1.5)	(1.5)	(1.5)
CM-Chitosan	(1.5)	(1.5)	(1.5)	1.5	(1.5)	(1.5)	(1.5)	(1.5)	(1.5)	(1.5)
O-Chitin	0.01	0.01	0.3	0.3	0.3	0.3	0.3	(0.3)	0.3	ND[e]

[a]Chitosan—solution in acetic acid; S-chitosan—chitosan sulphate; CM-chitosan—carboxymethyl chitosan; and O-chitin—chitin oligomers.

[b]Concentration (w/v) of chitin derivatives.

[c]Cmm—Clavibacter michiganense subsp. michiganense; Cmi—Clavibacter michiganense subsp. insidiosum; Ecp—Xanthomonas campestris pv. pelargonii; Xcph—Xanthomonas campestris pv. phaseoli; Psph—Pseudomonas syringae pv. phaseolicola; Pst—Pseudomonas syringae pv. tomato; Ea—Erwinia amylovora; Ecc—Erwinia carotovora subsp. carotovora; At—Agrobacterium tumefaciens; and EC—Escherichia coli.

[d]Parentheses indicate ineffectiveness at concentration of 1.5% (w/v).

[e]ND—not determined.

Table 2. Effect of chitin derivatives on hypersensitive
reaction (HR) of tobacco against Pseudomonas
syringae pv. phaseolicola.

Treatment[a]	Hypersensitive Reaction
Bacterium + chitosan	−
Bacterium + CM-chitosan	+
Bacterium + S-chitosan	+
Bacterium + water (control)	+
Chitosan + water	−

[a]Mixtures of bacteria at concentration of 5×10^7 with 0.05% chitin
derivatives were injected into Nicotiana tabacum cv. Xanthi nc leaves.
CM-chitosan—carboxymethyl chitosan.
S-chitosan—chitosan sulphate.
"+"—HR appeared.
"−"—HR did not appear.

The mechanism of this phenomenon is still unknown. A strong attachment of heterologous bacteria to the walls in tobacco leaves is essential to elicit the HR. Therefore, it is possible that chitin derivatives prevent the attachment of bacterial cells into the plant cell walls or affect their survival in the intercellular spaces.

ANTIVIRUS ACTIVITY OF CHITOSAN

Phaseolus vulgaris cvs Fana and Pinto, *Nicotiana tabacum* cvs Xanthi nc, and Samsun NN and *Chenopodium* amaranticolor were used as local lesion hosts. Plants of the same height, age, and vigor were used for each experiment. All the plants were grown in a glasshouse at 20–25°C for a 16-h photoperiod. Tested plants were mechanically inoculated with viruses using carborundum as an abrasive. The effect of chitosan on virus infection was calculated as the percentage reduction in the number of lesions produced by viruses on the chitosan-treated leaves in comparison with the control. To evaluate the validity of the data Student's *t*-test was used. Samples of chitosan with different degrees of polymerization and deacetylation were obtained as described earlier. Deaminated chitosan was obtained by nitrous acid treatment [12].

As shown previously [6–9] chitosan applied by spraying or inoculation on the leaves protected the various plant species against local and systemic infection caused by mechanically transmitted viruses. The chitosan efficiency in the inhibition of virus infection depended on the plant-virus system tested, mode of application, and dose of chitosan.

The aims of this study were to explain the role of the host in chitosan-induced resistance and to determine the role of physicochemical characteristics in its antiviral activity. Results presented in Table 3 show that the inhibition of virus infection by chitosan depended on the host type. The infection of tobacco virus (TMV) was completely inhibited in bean plants with a mixture of chitosan and TMV. At the same time no inhibition was observed on the tobacco plants. A similar trend was observed when chitosan was applied to different hosts 24 h before inoculation. The results obtained with alfalfa mosaic virus confirmed this phenomenon. Thus the results indicated that chitosan does not cause irreversible inactivation when present in inocula and that chitosan exerts its inhibitory effect via a host.

The efficiency of chitosan originated from shrimp, crab, and krill shells was compared with efficiency of low molecular fractions isolated from these samples as well as with the efficiency of monomers (Table 4). Our data show that the efficiency of chitosan antiviral activity was

Table 3. Effect of the antivirus activity of chitosan on the plant type.

Virus/Host	Inhibition in Inoculum[a]	(%) Spraying[b]
TMV		
Nicotiana tabacum		
Xanthi nc	0	49.8
Nicotiana tabacum		
Samsun NN	0	39.8
Chenopodium		
amaranticolor	86.8	63.5
Phaseolus vulgaris		
Pinto	100.0	88.1
AIMV		
N. tabacum Samsun NN	54.1	47.2
Chenopodium		
amaranticolor	69.9	54.7
Phaseolus vulgaris		
Fana	100.0	98.8

[a]10 min before inoculation virus was incubated in 0.1% solution of chitosan.
[b]Plants were sprayed with 0.1% solution of chitosan and 1 day later were inoculated with virus.
Viruses: TMV—tobacco mosaic virus; AIMV—alfalfa mosaic virus.

Table 4. Inhibition of alfalfa mosaic virus (AIMV) infection by chitosan with different degrees of polymerization.[a]

Inhibitor	Average Degree of Polymerization	Concentration (wt%)		
		0.1	0.01	0.001
N-Acetyl-D-glucosamine	1	15	0	ND
D-Glucosamine	1	50	10	ND
Crab chitosan	15	N	N	N
Crab chitosan	40	ND	100	96
Crab chitosan	>100	100	100	98
Shrimp chitosan	15	N	N	N
Shrimp chitosan	>100	100	100	97
Krill chitosan	15	100	98	40
Krill chitosan	100	100	100	82
Krill chitosan	300	100	100	78

[a]Reduction of local lesions production by AIMV. Bean plants were inoculated with virus a day after treatment with chitosan.
ND—not determined.
N—after treatment on plants appeared virus-like local necrotic lesions.

slightly increased with the degree of polymerization growth. A rather high efficiency of D-glucosamine monomer at concentration of 0.1 wt% should be emphasized (Table 4). However, the increase in D-glucosamine concentration up to 0.5 wt% did not improve its antiviral activity. The antiviral efficiency of chitosan with different deacetylation degrees is presented in Table 5.

It can be seen that the antiviral activity of chitosan was not significantly dependent on the degree of deacetylation. It should be noted that N-acetyl glucosamine, in contrast to glucosamine, practically had no antiviral activity. Table 6 shows that the deaminated chitosan

Table 5. Inhibition of alfalfa mosaic virus (AIMV) infection by chitosan.[a]

Inhibitor	Degree of Deacetylation (%)	Concentration (wt%)		
		0.1	0.01	0.001
Crab chitosan	>90	100	98	98
Crab chitosan	60–65	100	99	98
Shrimp chitosan	97–98	100	96	98
Shrimp chitosan	64	100	98	97
Krill chitosan	70	100	99	95
Chitin	0	50	0	0

[a]Reduction of local lesions production by AIMV. Bean plants were inoculated with AIMV a day after treatment with chitosan. Chitosan with DP > 100.

Table 6. Inhibition of alfalfa mosaic virus (AIMV) infection by chitosan with different degrees of deamination.[a]

Inhibitor	Number of Free Amino Groups (%)	Degree of Deamination, % to Control	Concentration (wt%)		
			0.1	0.01	0.001
Shrimp chitosan[b]	80	—	97	91	80
DA-1	65	15	100	99	86
DA-2	50	30	ND	100	98
DA-3	30	50	ND	100	99

[a]Reduction of local lesion production by AIMV on bean leaves. Bean plants were inoculated with AIMV a day after treatment with chitosan.
[b]Shrimp chitosan sample at DP > 100 and number of free amino groups of 80% was subject to deamination and correspondingly used as control.
ND—not determined.

samples were somewhat more effective especially at lower concentrations. This effect was more visible when another test system was used.

This system was based on the local lesion formation in the leaves of *N. tabacum* var. Samsun NN and Xanthi nc inoculated with TMV. It was shown earlier that these plants were less susceptible to the antiviral action of chitosan. Therefore, it could be expected that the differences in the activity of deaminated and original forms of chitosan will be more pronounced. The data presented in Table 7 support a suggestion that the deaminated chitosan oligomers inhibited TMV more effectively than the nonmodified polymeric chitosan.

The leaves of 8- to 9-week-old tobacco plants were sprayed with 0.03 wt% inhibitor in 0.015 wt% acetic acid solution a day prior to inocula-

Table 7. Antiviral activity of the original and deaminated chitosan.[a]

Inhibitor	Tobacco var.	Number of Lesions	Inhibition (%)
DA-2	Samsun NN	90	49
Shrimp chitosan		146	18
Control		178	—
DA-2	Xanthi nc	46	41
Shrimp chitosan		74	5
Control		78	—

[a]Reduction of local lesions production by TMV on the leaves of *Nicotiana tabacum* var. Samsun NN and *N. tabacum* var. Xanthi nc.

tion and then they were mechanically inoculated with TMV. The leaves of control plants were sprayed with 0.015 wt% acetic acid only. Local lesions were counted after 3 days. In every experiment from thirty-one to thirty-nine leaves were inoculated. Average data of two separate experiments are presented in Table 7.

ANTIVIROID ACTIVITY OF CHITOSAN

In our experiments a severe strain of potato spindle tuber viroid (PSTV) was used. The viroid was multiplied and maintained in tomato (*Lycopersicon esculentum*) cv. Rutgers. The inoculum was prepared by grinding the infected leaves in a precooled mortar and pestle, with 0.05 M K_2HPO_4 solution. Inoculations for bioassay were done by mechanically rubbing the tomato plants at the second leaf stage previously dusted with carborundum. Two weeks after inoculation the plants were trimmed and assessed for viroid symptoms in the lateral shoots. If the plants were not showing any symptoms, then back inoculations were given to the tomato seedlings cv. Rutgers. All experiments were done in a glasshouse at 25–30°C with a light intensity of approximately 6,000 Lx produced by sodium lamps for 18 h per day.

Viroids are the smallest, autonomously replicating plant pathogens. A potato spindle tuber viroid (PSTV) is easily transmitted in the field by workers or their tools contaminated with viroid. In this section we report our attempts to prevent the mechanical transmission of PSTV by chitosan. Tested PSTV containing extracts were incubated with different concentrations of chitosan, and these viroid-chitosan mixtures were used for inoculation of the tomato plants. Chitosan was added to the inoculum 10 minutes before inoculation.

As shown in Table 8 the infectivity of PSTV was almost completely inhibited by chitosan even at a concentration of 0.01 mg/ml. A dilute, 0.1 wt% solution of chitosan applied by spraying also effectively protected the tomato plants against viroid infection (Table 9). Chitosan applied several days prior to an inoculation was still found to be effective in reducing by 50–75% the number of tomato plants infected with PSTV.

We showed that chitosan also protected against PSTV infection in untreated parts of the chitosan-treated plants. The present study confirmed additionally the evidence that chitosan can be a potential inducer of the natural disease resistance; therefore, this polymer is responsible for control of the plant pathogens.

ANTIPHAGE ACTIVITY OF CHITOSAN

In our experiments the following bacteria and phages were used: gram-negative *Escherichia coli* strain B and its phages T-2 and T-7;

Table 8. Effect of various concentrations of chitosan on the infectivity of PSTV.

Concentration of Chitosan (wt%)	Tests Number of Plants Infected/Inoculated				
	1	2	3	4	5
0.0	4/6	5/6	4/6	7/7	7/7
0.001	1/6	3/6	1/6	3/7	4/7
0.01	0/6	1/6	0/6	1/7	1/7
0.1	0/6	0/6	0/6	0/7	0/7

Table 9. Effect of chitosan on the infection of tomato plants by PSTV.

Experiments[a]	Tests[b]				
	1	2	3	4	5
Rubbing with inoculum					
Control	5/6	5/7	6/6	7/7	6/7
Chitosan	0/6	0/7	2/7	2/7	0/7
Dry inoculation					
Control	6/6	6/6	6/6	1/6	—
Chitosan	1/6	0/6	3/6	1/6	—

[a]Tomato plants were sprayed with 0.1% solution of chitosan and 1 day later were inoculated mechanically: (1) by rubbing with inoculum prepared in 0.05 M K_2HPO_4 diluted 100 or 500 times (w/v); (2) by direct rubbing with infected leaves ("dry inoculation").
[b]Number of plants infected/inoculated.

gram-positive *Bacillus thuringensis* subsp. galleriae I-97 and its phages I-97A and I-97B; and lysogenic strain of *B. thuringensis* subsp. galleriae I-97. Two ml of the night culture of bacteria (on B medium) was added to 40 ml of M9 medium and incubated for 4 h at 30°C on a rotory shaker. A suspension of the chitosan solution at final concentrations of 0.01 or 0.001 wt% was added to the bacteria. The phages were added to the chitosan-treated bacteria and control samples (untreated), and then all samples were incubated for 16 h. After that, all samples were centrifuged, and the concentration of phages in supernatants was determined by the overlay method. In the case of lysogenic strain of *B. thuringensis* subsp. galleriae I-97, 2 ml of the night culture was added to 40 ml of M9 medium, and the bacteria cells were grown for 4 h at 30°C on the rotory shaker. Bacteria cells were treated with chitosan at final concentrations of 0.01 or 0.001 wt% as well as incubated for 2 days. All samples were centrifuged, and the concentration of phages in supernatants was determined by the overlay method.

Chitosan at a concentration of 0.01 and 0.001 wt% did not inhibit growth of bacteria *E. coli* and *B. thuringensis* subsp. galleriae I-97. Chitosan, at a concentration of 0.01 or 0.001 wt% added to suspension of *E. coli,* almost completely protected bacteria against infection from both phages T-2 and T-7 (Table 10). Similarly, chitosan very effectively inhibited the infection of *B. thuringensis* subsp. galleriae from both phages I-97A and I-97B (Table 11).

It is very interesting that chitosan also blocked a lysis of lysogenic strain of *Bacillus thuringensis* subsp. galleriae I-97 (Table 12). Chitosan, due to its antiphage activity, is proposed to be used in agriculture for the protection of *Rhizobium* sp. against its phages. As far as we know this is the first report on inhibition of phages infection by chitosan.

BIOSTIMULATION OF PLANT GROWTH BY CHITOSAN

Some evidence of the biostimulation of plant growth is presented on a base of seed germination. A correlation between the structure of chi-

Table 10. *Effect of chitosan on infection of* Escherichia coli *by phages.*

Concentration of Chitosan (wt%)	Inhibition of Infection (%)	
	Phage T-7	Phage T-2
0.0	0.0	0.0
0.001	99.9	99.8
0.01	99.9	100.0

Table 11. Effect of chitosan on infection of
Bacillus thuringiensis *ssp. galleriae I-97 by phages.*

Concentration of Chitosan (wt%)	Inhibition of Infection (%)	
	Phage I-97A	Phage I-97B
0.0	0.0	0.0
0.001	100.0	83.4
0.01	100.0	98.0

Table 12. Inhibition of spontaneously induced infection of phages in Bacillus thuringiensis *ssp. galleriae I-97 by chitosan.*

Concentration of Chitosan (wt%)	Inhibition of Infection (%)
0.0	0.0
0.001	99.6
0.01	100.0

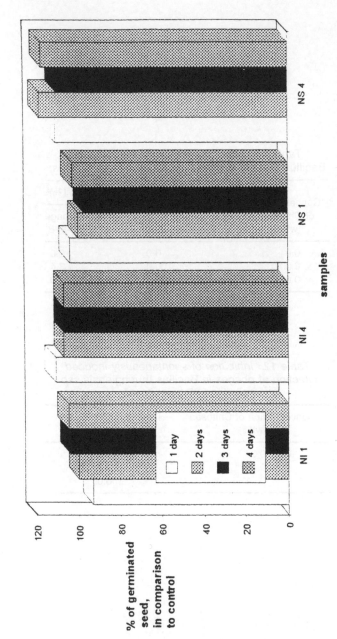

Figure 1. Effect of different type of coating agent on germination effectivity on wheat seed.

tosan and its bioactivity resulting from the amount and type of oligoaminosaccharides produced during the biodegradation process of this natural polymer plays a most important position in the antimicrobial activity of chitosan. At the same time, the oligoaminosaccharides obtained by chitosan biodegradation act on the biostimulation of plant growth. The biostimulation of wheat seed germination by different forms of chitosan coded as NI with average molecular weight of $\sqrt{\overline{M}_v} = 1 \times 10^5$, NS with average molecular weight of $\sqrt{\overline{M}_v} = 3 \times 10^5$, "1" with pH = 6.0 and "4" with pH = 7.0, used with 0.014 wt% for seed coating is presented in Figure 1. The results confirm that a lower molecular weight chitosan is characterized by a higher biostimulation effect.

CONCLUSIONS

The usefulness of chitin and its derivatives in agricultural applications is in the following areas: biostimulation of plant growth; plant protection against fungi, bacterias, and viruses; and postharvest protection. Biostimulation of the plant growth is mainly realized by seed coating, roots deepening, and plant dressing; whereas plant protection against fungi, bacteria, or viruses can be achieved by seed coating, plant dressing, and postharvest treatment. A special interest for the

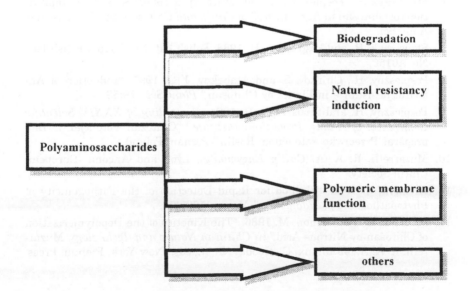

Figure 2. Polyaminosaccharides bioactivity phenomena.

use of chitin and its derivatives is postharvest treatment subjected to harvest protection (especially against fungi) and harvest freshness protection.

Bioactivity, a most important property of polyaminosaccharides, is a function of several phenomena (Figure 2). We can speculate that the antimicrobial activity of polyaminosaccharides relates mainly to the biodegradation phenomenon followed by stimulation of natural resistance, whereas antivirus activity depends on membrane and elicitor phenomena.

REFERENCES

1. Muzarrelli, R. A. A. 1978. *Chitin,* New York, Pergamon Press.
2. Roberts, A. F. R. 1992. *Chitin Chemistry,* Houndmills, Macmillan Press Ltd.
3. Rayan, C. A. "Oligosaccharides as Recognition Signal for the Expression of Defensive Genes in Plants," *Biochemistry,* 27:63.
4. Allan, C. R. and Hardwiger, L. A. 1979. "The Fungicidal Effect of Chitosan on Fungi of Varying Cell Wall Composition," *Exp. Mycol.,* 3:285.
5. Behamova, N. and Theriault, G. 1992. "Treatment with Chitosan Enhances Resistances of Tomato Plants to the Crown and Root Rot Pathogen *Fusarium oxysporum* F. sp. radicis-lycopersici," *Physiol. Mol. Plant Pathol.,* 41:33.
6. Struszczyk, H., Pospieszny, H. and Kotlinski, S. 1989. "Some New Applications of Chitosan in Agriculture," in *Chitin and Chitosan,* London, Elsevier Appl. Sci.
7. Struszczyk, H. and Pospieszny, H. 1988. Polish Pat. No. A0154, Polish Pat. No. A0156.
8. Pospieszny, H., Chirkov, S. and Atabekov, J. G. 1991. "Induction of Antiviral Resistance in Plants by Chitosan," *Plant Sci.,* 79:63.
9. Pospieszny, H. and Struszczyk, H. 1994. *Proceeding of XXXIV Scientific Conference of Plant Protection Institute,* "Chitozan–Potencjalny Biopreparat Przeciwko Patogenom Roslin," Poznan, Poland.
10. Muzarrelli, R. A. A. *Chitin Enzymology,* Lyon and Ancona, European Chitin Society.
11. Klement, Z. 1963. "Method for Rapid Detection of the Pathogenicity of Phytopathogenic *pseudomonas,*" *Nature,* 199:299.
12. Allan, G. G. and Peynon, M. 1986. "The Kinetics of the Depolymerization of Chitosan by Nitrous Acid," in *Chitin in Nature and Technology,* Muzarrelli, R. A. A., Jauniaux, C., Gooday, G. W., eds., New York, Plenum Press.

CHAPTER 12

Chitinolytic Enzymes in Selected Species of Veterinary Importance

MICHAEL LONDERSHAUSEN AND ANDREAS TURBERG
Institute for Parasitology
Bayer AG
51368 Leverkusen, Germany

INTRODUCTION

The degradation of chitin is a vital step in the life cycles not only of arthropods [1] but also during the development of parasitic nematodes [2]. In insects chitinases are important enzymes involved in molting. Excess, deficiency, or inhibition of enzyme activity may cause developmental failures. Two classes of enzymes catalyze the metabolism of chitin in insects: N-acetyl-β-D-glucosaminidase (exochitinase, EC.3.2.1.30) and endo-β-(1–4)-poly-N-acetylglucosaminidase (endochitinase, EC.3.2.1.14). Chitinases or chitinase-like molecules have also been reported in nematode eggs and larvae [3] as well as in microfilariae [4–6], indicating a crucial role not only in the hatching process but also during later steps of development within the host animal. To survive to its reproductive stage, hatching and each molt must be accomplished successfully by arthropod and nematode parasites. It is, therefore, of practical and general interest to further characterize parasite chitinases with regard to their heterogeneous expression during development as well as their susceptibility towards specific inhibitors, which might lead to new antagonizing agents useful in parasite control.

METHODS IN INVESTIGATION OF CHITINOLYTIC ENZYMES

Chitinolytic enzymes are secreted into the culture medium by the

185

epithelial insect cell line from *Chironomus tentans,* which was cultured essentially as described by Wyss [7]. For the investigation of arthropod parasites *Lucilia cuprina* developmental stages were selected from synchronized cultures (± 5 h) and kept in plastic dishes at 26–27°C and 70% relative humidity. Animals were staged according to Filshie [8]. Eggs from fully engorged female *Boophilus microplus* ticks (Parkhurst strain) were collected and kept at 26°C in the dark at 75% relative humidity. Eggs from days 10 and 17, as well as hatching larvae, unfed larvae, and fully engorged adults were used in the investigations. Eggs from the nematode *Haemonchus contortus* were purified from faeces according to Roberts and O'Sullivan [9] and hatched for L3 larvae. Adults of both sexes were collected from freshly killed sheep abomasum and used directly in the chitinolytic assays.

The endochitinase activity of parasite stages was determined as described previously. In brief, chitosan was acetylated with tritiated acetic anhydride by a modification of the method of Molano et al. [10]. The specific activity of the radiolabeled chitin was obtained by measuring the radioactivity after enzymatic hydrolysis. A specific activity of 1.07 mCi/g chitin was achieved. The tritium-labeled chitin was used as a substrate, and TCAsp of the chitinase reaction was measured after centrifugation in a Packard scintillation counter with an efficiency of 45%. Exochinase activity of *Lucilia cuprina* was determined by measuring the liberation of *p*NP from *p*NP-GlcNAc. Fifty microliters of 0.2 M citrate phosphate buffer (pH 5.5), with the appropriate substrate concentration, and 50 μl homogenate in the same buffer were mixed and incubated at 25°C for 10 minutes and then stopped by addition of 700 μl 0.01 M NaOH. The concentration of the phenolate anion was measured at 410 nm.

Chitinolytic activity after column chromatography was determined with a highly sensitive fluorescence assay slightly modified from McCreath and Gooday [11]. Fifty microliters of diluted extract and 50 μl of MUF-chitotriose (80 μM) were incubated for 10–15 minutes at 20°C. Fluorescence of released MUF was measured in a Kontron spectral fluorometer at 350 nm excitation and 440 nm emission wavelength after enhancement of fluorescence at pH 11 (100 mM sodium hydroxide/glycine). For comparison, incubation conditions were maintained constant within species. Samples from parasites were prepared by homogenization in sodium phosphate buffer (0.1 M, pH 7.0, 100 g·l⁻¹ glycerol) with an all-glass potter homogenizer (*Lucilia, Chironomus*) or in a liquid nitrogen cooled mortar (*Boophilus, Haemonchus*). All homogenates were subsequently treated with ultrasonics (3 × 5 sec, 50 W) and centrifuged for 1 h at 100,000 × g at 4°C. Aliquots of supernatants were diluted to appropriate protein concentrations and either stored at $-25\,^{\circ}$C or directly used for enzyme assays. Enzyme activity in stored extracts was stable for at least 6 months. Column chromatography was carried out on Beckman Ultra Spherogel™-SEC 3000 in sodium phosphate buffer (0.1 M, pH 7.0) at 0.5 ml/min flow rate on a Waters HPLC system. Fractions (100-μl) were collected from void to column volume and tested as described for enzymatic assays. Protein content was determined according to Bradford [12] using bovine serum albumin as standard.

For statistical analysis the K_M values were computed by rearrangement of the data according to Lineweaver-Burk [13]. For determination of IC_{50} values, a logit/log transformation was performed. The data were fit to straight lines by the linear least squares method, and standard error values were calculated [14]. For inhibitor investigations the compounds dissolved in 5 μl DMSO were added. Controls revealed no effects of this solvent on the enzymatic activities.

CHARACTERIZATION OF *LUCILIA CUPRINA* CHITINASE

The availability of tritiated chitin allows a simple assay, based on the fact that the enzymatic reaction products of endochitinase [10,15] are soluble in aqueous TCA, whereas the remaining substrate chitin is not. To characterize the chitinases of *Lucilia cuprina*, an investigation was conducted covering the determination of kinetic and some physicochemical parameters. Chitinase activity was highest in 10-day-old pupae of *Lucilia cuprina* with an activity of 587 $\pm$ 43 pg TCAsp/h/mg protein, which corresponds to 212% of the activity of 24-h larvae (100%). With radiolabeled chitin as substrate, *Lucilia* pupae endochitinase exhibited typical Michaelis-Menten kinetics (Figure 1) with an apparent K_M value of 2.05 $\pm$ 0.25 mg chitin (95% Cl, $R = 0.98$) and a V_{max} of 774 pg TCAsp/h/mg protein. Comparable results were achieved for *Chironomus tentans* chitinase with an apparent K_M value of 1.5 $\pm$ 0.2 mg chitin (95% Cl, $R = 0.97$) and a V_{max} of 1,500 pg TCAsp/h/mg protein. At incubation times longer than 60 minutes a loss of enzymatic activity was observed (Figure 1, inset). Therefore no incubation times exceeding 1 h were used in chitinolytic assays.

The assay of endochitinase described in this report is simple and extremely sensitive. In those experiments in which the decomposition of chitin was measured within 60 minutes (Figure 1, inset) a linear relationship between enzymatic activity and incubation time was observed. Possible explanations for the loss of linearity at longer incubation times are the inhibition by reaction products or heterogeneity of the chitin substrate [10].

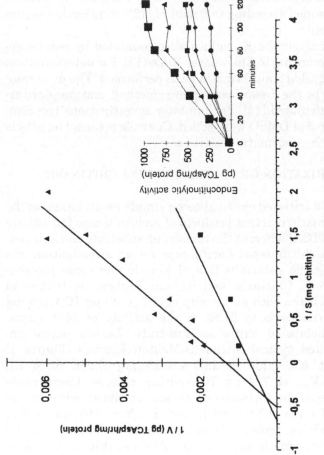

Figure 1. Determination of kinetic properties of chitinases from *Lucilia cuprina* (▲) and *Chironomus tentans* cells (■). The data are given as a Lineweaver-Burk plot. Each point represents the average of four determinations. Inset: Time curves for TCA-soluble products generated by degradation of [³H]-chitin by *Lucilia cuprina* chitinase. Reactions were initiated by adding the biological sample to standard reaction mixtures containing different amounts of chitin. Each point shown represents separate determinations carried out in triplicate. Substrate concentrations were: 4.05 mg (■), 3.15 mg (▲), 2.25 mg (■), 1.35 mg (♦), and 0.45 mg (●) [³H]-chitin.

The products of the enzymatic reaction chitobiose and chitotriose inhibit endochitinolytic activity (Table 1). However, an approximate calculation of expected inhibition values caused by the production of di- and trimers of GlcNAc showed that only a negligible part of the deviation from linearity, observed between 80 and 120 minutes, could be attributed to product inhibition. Similarly, a time-dependent loss of enzymatic activity does not seem to be the reason for the drop in reaction rate [10,15]. Finally, one should consider the possibility that chitin could behave as a heterogeneous substrate [10], influencing the degradation velocities of core and surface of the chitin particles in different quantities. Theoretical kinetic models [15,16] that best fit experimental data for polysaccharide degradation include the formation of active and nonactive enzyme-chitin complexes, as well as the generation of short intermediate oligosaccharides [16] by endochitinase, which are the best substrates for hydrolysis by exochitinase [17]. In this context exo-β-(1–4)-N-acetylglucosaminidase was tested using pNP-GlcNAc as substrate, to obtain a complete picture of chitin degradation in *Lucilia cuprina*. In accordance with data from Reference [18] chitinase was inhibited only to 5 ± 2% by 0.24 mM acetamido-galactonolactone, a concentration that is sufficient to totally block exochitinase from *Lucilia cuprina*. The corresponding K_M values for the chitinolytic enzymes of *Lucilia cuprina* and *Chironomus tentans* were similar to those described for the isolated enzymes of *Drosophilia melanogaster* Kc-cells

Table 1. Inhibitory effects of different substances on chitinolytic activity of Lucilia cuprina *chitinase.*

Substance	IC$_{50}$ (mM)
GlcNAc	>10
Chitobiose	4.0
Chitotriose	1.4
Chitotetraose	1.2
Acetamido-galactonolactone	6.0
Cinerubin B	0.6
Tunicamycin[a]	1.0
RBN 126	1.3
Plumbagin	1.4
Melanin[b]	280 μg
Dinactin	2.7
Avermectin B1	>10
Triflumuron	>10

[a]Mixture of Tunicamycin: A (5%), B (34%), C (40%), and D (18%), mean weights were used for calculation.
[b]μg are given, because of weight distribution of melanin.

(apparent K_M = 2.2 mg chitin) and *Artemia salina* nauplii (apparent K_M = 1.8 mg chitin [19,20]. Summarizing our data and those described for insect chitin degradation [1], it seems that the chitinolytic system of *Lucilia cuprina* consists of at least two classes of enzymes that successively break down chitin to yield GlcNAc. Endochitinase acts upon chitin and exochitinase further hydrolyzes the products of endochitinase [21].

To get indications of the amino acids that might be of importance for enzymatic activity, the pH dependency [Figure 2(a)] was determined. At those pH values (pH 5.3 and pH 8.3) where the activity dropped precipitously the activation enthalpies [Figure 2(b)] were measured. The corresponding enthalpies were 6.5 ± 1.3 kcal/mol (pH 5.3, R = 0.97, 95% Cl) and 5.17 ± 1.1 kcal/mol (pH 8.3, R = 0.99, 95% Cl), respectively. The pH-dependent activity studies of *Lucilia cuprina* chitinase revealed at least two ionizable groups at apparent pK_a's of 5.3 and 8.3, determined according to Bisswanger [22].

Taking into account the sometimes enormous differences in pK_a's of free amino acids and those that are part of a protein molecule [22], one could speculate on the participation of an acidic group [23], which may be an aspartic acid or glutamic acid residue in the active center, as is the case in lysozyme also [24]. This is in accordance with the evidence of several invariant aspartic and glutamic acid residues in the presumed active site of different polysaccharidases [25], including the first cloned chitinase from an insect, *Manduca sexta* [26]. Although lysozyme exhibits a relatively narrow pH optimum of pH 4–6, *Lucilia cuprina* endochitinase has an extended pH optimum into the alkaline pH range. Likely candidates for groups ionizing at pH 8.3 are imidazole, sulfhydryl or amino side chains. None of the corresponding activation enthalpies determined for amino acids were in the range of the determined value of about 5 kcal/mol.

CHITINASE INHIBITORS

To investigate substances interfering with chitin degradation, some naturally occurring or synthetic compounds that are supposed to have an interrelationship to chitin metabolism or inhibition of invertebrate development were examined for their inhibitory potency in the introduced test systems.

It has been observed that few compounds belonging to the group of polyketide antibiotics (cinerubin B), amino sugars (RBN 126), or substances that are related to the mechanism of cuticular buildup (melanization, sclerotization), and glycoprotein synthesis (tunicamycin) shared a more or less good inhibitory effect (Table 1, Figure 4).

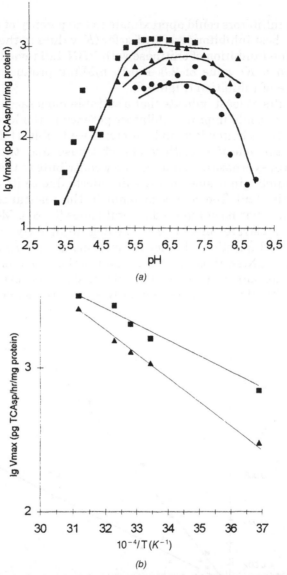

Figure 2. Determination of pH dependency and effect of temperature on chitinolytic activity. (a) pH profiles of chitinolytic activity measured in citrate buffer (■: pH 2.75–7.25, 100 mM); sodium phosphate buffer (▲: pH 5.0–8.5, 100 mM); and acetate phosphate buffer (●: pH 5.5–9.0). Concentration of [³H]-chitin was 7 mg per assay. Mean values from four determinations are shown. pH values were controlled prior to and after incubation. (b) Dependency of chitinolytic activity from temperature (Arrhenius plot) at pH 5.3 (▲) and pH 8.3 (■). Other conditions are as described under Figure 2(a).

None of the inhibitors could approximate to the potency of allosamidin, which is the best inhibitor described so far (K_i values in the range of 0.1 μM, [27]). The inhibition of chitinase with RBN 126 revealed an IC_{50} of 1.3 mM and a K_i value of 2.8 ± 0.16 mM/mg protein with a competitive type of inhibition (Figure 3).

These results at least indicate that a suitable monosaccharide substitution pattern could improve inhibitory potency to the level of the preferred substrates chitotriose and chitotetraose (Table 1). Melanin and plumbagin are effective against endochitinase and the latter also against chitin synthase [28], and this may contribute to the insecticidal action of plumbagin besides nonspecific effects due to its electron carrier properties [29]. The role of melanin in the resistance of fungi to microbial lysis has been reported several times [30,31]. Melanin binds both to the substrate chitin and to the degrading enzyme, posing the still unresolved question whether it functions as an enzyme inhibitor or simply as a physical barrier that prevents the access of chitinase to its appropriate substrate. Although melanin does not reveal any insecticidal activity, this substance generally shares the same mode of ac-

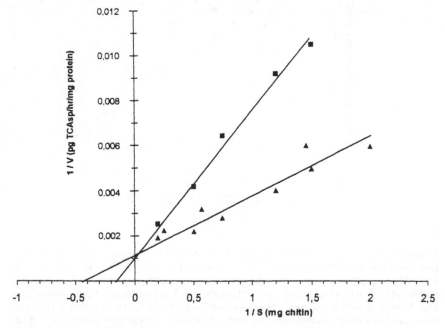

Figure 3. Determination of kinetic properties of *Lucilia cuprina* chitinase in the absence (▲) and presence of 14 mM RBN 126 (■). Each point represents the average of four measurements.

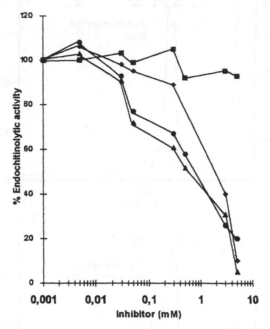

Figure 4. Effect of different inhibitors on chitinolytic activity. Each point shown represents the mean value of three determinations. 100% corresponds to 630 pg TCAsp/h/mg protein. Cinerubin B (▲), tunicamycin (●), RBN 126 (♦), and GlcNAc (■).

tion for microbial and insecticidal polysaccharase inhibition, perhaps stimulating new ideas on influencing insect development. Inhibition of chitin metabolism in *Mucor miehei* and *Artemia salina* by ivermectin has also been reported [32]. In the light of the proposed mode of action of ivermectin as effector of glutamate gated chloride channels [33], it now becomes less likely that this fermentation product is able to kill susceptible organisms also by inhibiting chitin turnover and synthesis. No inhibition or stimulation could be obtained with triflumuron as it was originally presumed by Deul et al. [34] for benzoylphenylureas (Figure 4).

MOLECULAR WEIGHT PATTERN OF CHITINASES FROM DIFFERENT VETERINARY PARASITES

In addition to these investigations the molecular weight patterns of chitinases from different important classes of veterinary parasites and life stages were determined. As shown in Figure 5(a) to 5(g) chitinolytic enzymes could be demonstrated in all stages tested, but the activity

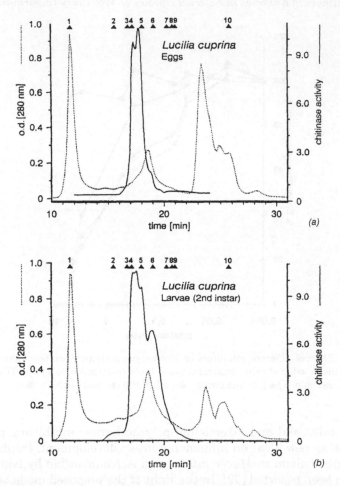

Figure 5. Enzyme activity profile after HPLC-chromatography of cytosolic extracts from different parasite developmental stages. Chitinase activity is given as relative fluorescence (rf) units; 1% rf ≅ 0.016 nmol MUF liberated within 30 minutes per mg protein. (a) HPLC-molecular sieve separation of cytosolic extracts from *Lucilia cuprina* eggs. (b) HPLC-molecular sieve separation of cytosolic extracts from *Lucilia cuprina* second larval instar. (c) HPLC-molecular sieve separation of cytosolic extracts from *Lucilia cuprina* adults. (d) HPLC-molecular sieve separation of cytosolic extracts from *Boophilus microplus* eggs (10 d). (e) HPLC-molecular sieve separation of cytosolic extracts from *Boophilus microplus* eggs (17 d). (f) HPLC-molecular sieve separation of cytosolic extracts from *Boophilus microplus* adults. (g) HPLC-molecular sieve separation of cytosolic extracts from *Haemonchus contortus* larvae. Closed triangles indicate standard protein retention times (1: void volume, 2: ferritin, 3: aldolase, 4: phosphorylase b, 5: bovine serum albumin, 6: ovalbumin, 7: carboanhydrase, 8: chymotrysinogen a, 9: cytochrome C, 10: column volume). Molecular weights were calculated from linear regression curves from standard proteins.

194

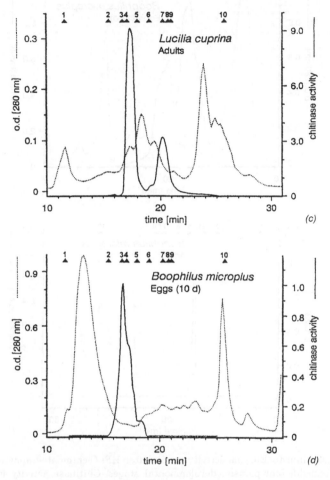

Figure 5 (continued). Enzyme activity profile after HPLC-chromatography of cytosolic extracts from different parasite developmental stages. Chitinase activity is given as relative fluorescence (rf) units; 1% rf $\cong$ 0.016 nmol MUF liberated within 30 minutes per mg protein. (c) HPLC-molecular sieve separation of cytosolic extracts from *Lucilia cuprina* adults. (d) HPLC-molecular sieve separation of cytosolic extracts from *Boophilus microplus* eggs (10 d). Closed triangles indicate standard protein retention times (1: void volume, 2: ferritin, 3: aldolase, 4: phosphorylase b, 5: bovine serum albumin, 6: ovalbumin, 7: carboanhydrase, 8: chymotrysinogen a, 9: cytochrome C, 10: column volume). Molecular weights were calculated from linear regression curves from standard proteins.

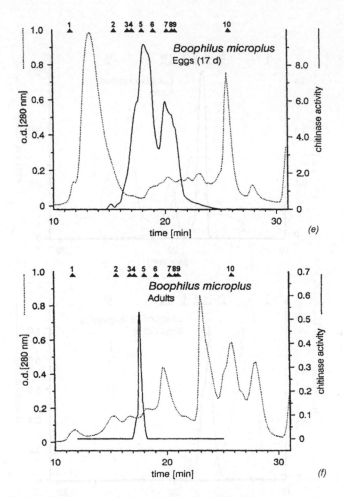

Figure 5 (continued). Enzyme activity profile after HPLC-chromatography of cytosolic extracts from different parasite developmental stages. Chitinase activity is given as relative fluorescence (rf) units; 1% rf ≅ 0.016 nmol MUF liberated within 30 minutes per mg protein. (a) HPLC-molecular sieve separation of cytosolic extracts from *Lucilia cuprina* eggs. (b) HPLC-molecular sieve separation of cytosolic extracts from *Lucilia cuprina* second larval instar. (c) HPLC-molecular sieve separation of cytosolic extracts from *Lucilia cuprina* adults. (d) HPLC-molecular sieve separation of cytosolic extracts from *Boophilus microplus* eggs (10 d). (e) HPLC-molecular sieve separation of cytosolic extracts from *Boophilus microplus* eggs (17 d). (f) HPLC-molecular sieve separation of cytosolic extracts from *Boophilus microplus* adults. (g) HPLC-molecular sieve separation of cytosolic extracts from *Haemonchus contortus* larvae. Closed triangles indicate standard protein retention times (1: void volume, 2: ferritin, 3: aldolase, 4: phosphorylase b, 5: bovine serum albumin, 6: ovalbumin, 7: carboanhydrase, 8: chymotrysinogen a, 9: cytochrome C, 10: column volume). Molecular weights were calculated from linear regression curves from standard proteins.

196

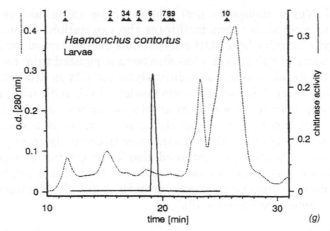

Figure 5 (continued). Enzyme activity profile after HPLC-chromatography of cytosolic extracts from different parasite developmental stages. Chitinase activity is given as relative fluorescence (rf) units; 1% rf ≅ 0.016 nmol MUF liberated within 30 minutes per mg protein. (g) HPLC-molecular sieve separation of cytosolic extracts from *Haemonchus contortus* larvae. Closed triangles indicate standard protein retention times (1: void volume, 2: ferritin, 3: aldolase, 4: phosphorylase b, 5: bovine serum albumin, 6: ovalbumin, 7: carboanhydrase, 8: chymotrysinogen a, 9: cytochrome C, 10: column volume). Molecular weights were calculated from linear regression curves from standard proteins.

and molecular pattern were different among the parasites. Strong chitinolytic activity could be demonstrated with the sensitive fluorometric endochitinase-specific assay in *Lucilia cuprina* eggs, second instar larvae, and adults in the region of 85 kDa. In addition, stage-specific chitinolytic activity was shown for second instar larvae at 50 kDa and for adults at 24 kDa.

A more pronounced stage-specific expression of distinct chitinases seems to occur in the tick *Boophilus microplus*. Developing eggs at day 10 had chitinolytic activity mainly in the region around 100 kDa. Shortly before larval hatching, 17-day-old eggs showed eight- to tenfold higher chitinolytic activity at about 87 kDa and 60 kDa. The comparably low-enzyme activity of adults could be demonstrated only at 78 kDa. Enzymes of this size could also contribute to the 75 kDa shoulder in 17-day-old eggs. In the nematode *Haemonchus contortus* one main peak at 37 kDa was detected. Chitinolytic activity was due to true endochitinase since lysozyme, which has some chitinolytic activity, could contribute to chitinolysis in *Lucilia cuprina* only for less than 2%. Lysozyme activity could not be detected in *Chironomus* cells. Profound changes in total chitinolytic activities between different developmental stages were also obtained in other arthropods such as *Ephestia cautella*

[35] and *Musca domestica* [36] and could be explained as a molt-preparing mechanism that facilitates the exuviation. Stimulation of chitinolytic activity by 20-OH-ecdysone, as demonstrated for *Chironomus tentans* cell cultures [37] has also been suggested from the correlation of ecdysteroid content and chitinolytic activity in different insects as *Bombyx mori* [38] and *Ephestia cautella* [35], indicating a causal relationship between insect ecdysteroid hormones and chitin-degrading enzymes, although the underlying mechanism of transcriptional control still has to be clarified. An increase in chitinolytic enzymes during premolt or hatching as obtained also for *Boophilus microplus* eggs and *Haemonchus contortus* larvae indicate that chitin might play a role not only during egg development in ticks but also during different stages of nematode life cycle [3].

SUMMARY

A test system for quantitative determination of inhibitory effects on chitin-metabolizing enzymes was established. The test system was used for the mechanistic examination of known and potential insecticides, referred to as their insect growth regulator activity.

Radiolabeled chitin, prepared by acetylation of chitosan with tritiated acetic anhydride, was used as substrate in a rapid and sensitive assay for endochitinase activity. Analyses of kinetic parameters revealed an apparent K_M value 2.05 mg chitin. Investigation of potential inhibitors revealed half-maximal inhibition at about 1 mM for cinerubin B, tunicamycin, plumbagin, and RBN 126. The amino sugar derivative RBN 126 inhibits chitinolytic activity in a competitive manner. In addition, melanin, a substance already described for its role in sclerotization, displayed an inhibitory effect on chitinase. Analysis of pH and temperature dependency of kinetic parameters suggested that at least two ionizable groups are involved in catalysis, with apparent pK_a's of 5.3 and 8.3. Determination of the activation energies revealed values of 6.5 kcal/mol at pH 5.3 and 5.17 kcal/mol at pH 8.3. Different developmental stages of insects and tick were demonstrated to express chitinolytic enzymes of different molecular weights.

ABBREVIATIONS

Acetamido-galactonolactone (2-acetamido-2-deoxy-D-galactonolactone); triflumuron (1-(4-trifluoromethoxyphenyl)-3-(2-chloro-benzoyl)-urea); chitobiose (2-acetamido-2-deoxy-4-*O*-(2-acetamido-2-deoxy-β-D-glucopyranosyl)-D-glucopyranose); chitotetraose (*N,N',N'',N'''*-tetraacetyl-β-chitotetraose); chitotriose (N,N',N''-triacety-β-chitotrioside); Cl (confidence limit); GlcNAc (2-acetamido-2-deoxy-D-glucose); IC_{50} (inhibitory concentration to suppress 50% of the initial activity); MUF (4-methylumbelliferyl); plumbagin (5-hydroxy-2-methyl-1,4-naphthoquinone); pNP (p-nitrophenyl); pNP-GlcNAc (p-nitrophenyl-N-acetyl-β-D-glucosaminide); RBN 126 (1-methoxy-2-(2,4-dinitro-anilino)-2-deoxy-α-D-glycopyranoside); TCA (trichloroacetic acid); TCAsp (TCA-soluble products).

ACKNOWLEDGEMENTS

We thank Iris Schröder and Sabine Geißreiter for skillful technical assistance and preparation of biological samples as well as Heidemarie Heim-Londershausen and Claudia Hüter for carefully writing the manuscript.

REFERENCES

1. Kramer, K. I., Dziadik-Turner, C. and Koga, D. 1985. "Chitin Metabolism in Insects," in *Comprehensive Insect Physiology, Biochemistry and Pharmacology, Vol. 3.* Academic Press, New York, eds. Kerkut, G. A. and Gilbert, L. J., pp. 75–115.

2. Cohen, E. 1993. "Chitin Synthesis and Degradation as Targets for Pesticide Action," *Arch. Insect Biochem. Physiol.,* 22:245–261.

3. Justus, D. E. and Ivey, M. H. 1969. "Chitinase Activity in Developmental Stages of *Ascaris suum* and Its Inhibition by Antibody," *J. Parasitol.,* 55(3):472–476.

4. Fuhrmann, J. A., Lane, W. S., Smith, R. F., Piessens, W. F. and Perler, F. B. 1992. "Transmission Blocking Antibodies Recognize Microfilarial Chitinase in Brugian Lymphatic Filariasis," *Proc. Natl. Acad. Sci. USA,* 89:1548–1552.

5. Gooday, G. W., Brydon, L. I. and Chappell, L. H. 1988. "Chitinase in Female *Onchocerca gibsoni* and Its Inhibition by Allosamidin," *Mol. Biochem. Parasitol.,* 29:223–225.

6. Raghavan, N., Freedman, D. O., Fitzgerald, P. C., Unnasch, T. R., Ottesen, E. A. and Nutman, T. B. 1964. "Cloning and Characterization of a Potentially Protective Chitinase-like Recombinant Antigen from *Wucheria bancrofti*," *Int. Immun.,* 62(5):1901–1908.

7. Wyss, C. 1982. "*Chironomus tentans* Epithelial Cell Lines Sensitive to Ecdysteroids, Juvenile Hormone, Insulin and Heat Shock," *Exp. Cell Res.,* 139:309–319.

8. Filshi, B. K. 1970. "The Fine Structure and Deposition of Larval Cuticle of the Sheep Blowfly (*Lucilia cuprina*)," *Tissue and Cell,* 2(3):479–498.

9. Roberts, F. H. S. and O'Sullivan, P. J. 1950. "Methods for Egg Counts and

Larval Cultures for Strongyles Infesting the Gastro-Intestinal Tract of Cattle," *Agric. Res.*, 1(1):99–102.

10. Molano, I., Duran, A. and Cabib, E. 1977. "A Rapid and Sensitive Assay for Chitinase Using Tritiated Chitin," *Anal. Biochem.*, 83:648–656.

11. McCreath, K. J. and Gooday, G. W. 1992. "A Rapid and Sensitive Microassay for the Determination of Chitinolytic Activity," *J. Microbiol. Meth.*, 14:229–237.

12. Bradford, M. 1976. "A Rapid and Sensitive Method for the Quantitation of Microgram Quantities of Protein Utilising the Principle of Protein Dye Binding," *Anal. Biochem.*, 72:248–254.

13. Lineweaver, H. and Burk, D. 1934. "The Determination of Enzyme Dissociation Constants," *J. Am. Chem. Soc.*, 56:658–666.

14. Linder, A. and Berchtold, W., eds. 1976. *Statistische Auswertung von Prozentzahlen: Probit- und Logit-Analyse mit EDV,* Birkhäuser Verlag, Basel, pp. 20–33.

15. Fukamizo, T. and Kramer, K. I. 1985. "Mechanism of Chitin Hydrolysis by the Binary Enzyme Chitinase System in Insect Moulting Fluid," *Insect Biochem.*, 15(1):141–145.

16. Koga, S., Mai, M. S., Dziadik-Turner, C. and Kramer, K. I. 1982. "Kinetics and Mechanism of Exochitinase and β-N-Acetylglucosaminidase from the Tobacco Hormone, *Manduca sexta* L.," *Insect Biochem.*, 12(5):493–499.

17. Koga, S., Ilka, I. and Kramer, K. J. 1983. "Insect Endochitinases: Glycoproteins from Moulting Fluid, Integument and Pupal Haemolymph of *Manduca sexta* L.," *Insect Biochem.*, 13(3):295–305.

18. Findlay, J., Levvy, G. A. and Marsh, C. A. 1958. "Inhibition of Glycosidases by Aldonolactones of Corresponding Configuration," *Biochem. J.*, 69:467–476.

19. Boden, N., Sommer, U. and Spindler, K.-D. 1985. "Demonstration and Characterization of Chitinases in the *Drosophilia* K_c Cell Line," *Insect Biochem.*, 15(1):19–23.

20. Funke, B. and Spindler, K.-D. 1989. "Characterization of Chitinase from the Brine Shrimp Artemia," *Comp. Biochem. Physiol.*, 94B(4):691–695.

21. Mommsen, T. P. 1980. "Chitinase and β-N-Acetylglucosaminidase from the Digestive Fluid of the Spider, *Cupiennius salei*," *Biochem. Biophys. Acta*, 612:361–372.

22. Bisswanger, H., ed. 1979. *Theorie und Methoden der Enzymkinetik*, Verlag Chemie, pp. 125–129.

23. Bedi, G. S., Shah, R. H. and Bahl, O. P. 1984. "Studies on Turbatrix Aceti-β-N-Acetylglucosaminidase," *Arch. Biochem. Biophys.*, 233(1):251–259.

24. Imoto, T., Johnson, N., North, A. C. T., Phillips, D. C. and Dupley, G. 1972. "Vertebrate Lysozymes," in *The Enzymes, Vol. 7.* Acad. Press, New York, ed. Boger, P. D., pp. 665–868.

25. Henrissat, B. 1990. "Weak Sequence Homologies among Chitinases Detected by Clustering Analysis," *Prot. Seq. Data Anal.*, 3:523–526.

26. Kramer, K. J., Corpuz, L., Choi, H. K. and Muthukrishnan, S. 1993. "Sequence of a cDNA and Expression of the Gene Encoding Epidermal and Gut Chitinases of *Manduca sexta*," *Insect Biochem. Mol. Biol.*, 23:691–701.

27. Koga, D., Isogai, A., Sakuda, S., Matsumoto, S., Suzuki, A., Kimura, S. and Ide, A. 1987. "Specific Inhibition of *Bombyx mori* Chitinase by Allosamidin," *Agric. Biol. Chem.*, 51:471–476.

28. Londershausen, M., Turberg, A., Buss, U., Spindler-Barth, M. and Spindler, K.-D. 1993. "Comparison of Chitin Synthesis from an Insect Cell Line and Embryonic Tick Tissues," in *Chitin Enzymology*, ed. Muzzarelli, R. A. A., *Eur. Chitin Soc.*, pp. 101–108.

29. Leicht, W. and Nauen, R. 1985. "Inhibition of Chitin Biosynthesis by Plumbagin," Bayer Research Account, ZF-FBT 12/49.

30. Bull, A. T. 1970. "Inhibition of Polysaccharases by Melanin:Enzyme Inhibition in Relation to Mycolysis," *Arch. Biochem. Biophys.*, 137:345–356.

31. Kuo, M. J. and Alexander, M. 1967. "Inhibition of the Lysis of Fungi by Melanins," *J. Bacteriol.*, 94:624.

32. Calcott, P. H. and Fatig, R. O. 1984. "Inhibition of Chitin Metabolism by Avermectin in Susceptible Organisms," *J. Antibiot.*, 37(3):253–259.

33. Cully, D. F., Vassilatis, D. K., Liu, K. K., Paress, P. S., Van der Ploeg, L. H. T., Schaeffer, J. M. and Arena, J. P. 1994. "Cloning of an Avermectin-sensitive Glutamate-Gated Chloride Channel from *Caenorhabditis elegans*," *Nature*, 371:707–710.

34. Deul, D. H., De Jong, B. J. and Kortenbach, J. A. M. 1978. "Inhibition of Chitin Synthesis by Two 1-(2,6-Disubstituted Benzoyl)-3-phenylurea Insecticides. II," *Pest. Biochem. Physiol.*, 8:98–105.

35. Spindler-Barth, M., Shaaya, E. and Spindler, K.-D. 1986. "The Level of Chitinolytic Enzymes and Ecdysteroids during Larval-Pupal Development in *Ephestia cautella* and the Modification by Juvenile Hormone Analogue," *Insect Biochem.*, 16(1):187–190.

36. Sing, G. I. P. and Vardamis, A. 1984. "Chitinases in the House Fly *Musca domestica*. Pattern of Activity in the Life Cycle and Preliminary Characterization," *Insect Biochem.*, 13:125–128.

37. Quack, S., Fretz, A., Spindler-Barth, M. and Spindler, K.-D. 1995. "Receptor Affinities and Biological Responses of Nonsteroidal Ecdysteroid Agonists on the Epithelial Cell Line from *Chironomus tentans* (Diptera: Chironomidae)," *Eur. J. Entomol.*, 92:341–347.

38. Kimura, S. 1973a. "The Control of Chitinase Activity by Ecdysterone in Larvae of *Bombyx mori*," *J. Insect Physiol.*, 19:115–123.

MEDICINE AND BIOTECHNOLOGY

Applications of Chitin, Chitosan, and Their Derivatives to Drug Carriers for Microparticulated or Conjugated Drug Delivery Systems

Hiraku Onishi, Tsuneji Nagai and Yoshiharu Machida

Hoshi University

Tokyo, Japan

INTRODUCTION

In the pharmaceutical field, many kinds of macromolecules have been used to control drug behavior in various dosage forms. Chitin is obtained as a natural product and chitosan is easily produced from chitin. Many derivatives of chitin and chitosan are synthesized. N-Succinyl-chitosan and 6-O-carboxymethyl-chitin are well-known as water-soluble polymers. These materials are supplied inexpensively. Furthermore, many of them show high biocompatibility [1] and biodegradability [2–5]. Therefore, chitin, chitosan, and their derivatives have been attractive as materials for medical and pharmaceutical use. Recently, for drug targeting and control of biopharmaceutical properties of drugs, microspheres, nanoparticles, liposomes, microemulsions, and macromolecular conjugates have been developed as part of drug delivery systems [6–14]. Biocompatible and biodegradable polymers are suitable for micro- or nanospheres or macromolecular conjugates. Thus, chitin, chitosan, and their derivatives are utilized for these drug delivery systems. Antitumor drugs are selected as agents to be controlled by the drug delivery systems because they exhibit toxic effects as well as therapeutic effects in many cases. These considera-

205

tions suggest that the combination of chitin, chitosan, and their derivatives with antitumor drugs should be useful for drug delivery systems of micro- or nanospheres or macromolecular conjugates. The following descriptions are based mainly on the results of examination of chitosan microspheres containing a drug (5-fluorouracil) [15,16], chitosan-drug (cytarabine) conjugate [17], N-succinyl-chitosan-drug (mitomycin C) conjugate [18–20], and 6-O-carboxymethyl-chitin-drug (mitomycin C) conjugate [18–20].

CHITOSAN MICROSPHERES

Preparation and Properties of Various Kinds of Chitosan Microspheres

Since chitosan is soluble in an acetic acid aqueous solution, a solvent evaporation process in an oil phase [21], (i.e., dry-in-oil method) can be utilized. Chitosan with a deacetylation degree of 63% was used. By using the dry-in-oil method, simple chitosan microspheres containing 5-fluorouracil (5-FU), MS-A, were prepared as follows: 5-FU was dissolved in 1.5% (w/v) chitosan solution, which was prepared using 2% acetic acid aqueous solution, to make up 0.5% (w/v) 5-FU solution. Fifty milliliters of the solution was dropped gradually into 500 ml of rapeseed oil under stirring at 300 rpm. The system was warmed to 50°C at a speed around 10°C/h, the pressure was reduced for 48 h by an aspirator, and the solvent was removed completely by a vacuum pump (5 h). During this operation, stirring was continued. After that, the microspheres produced were separated, washed successively with 10% NaOH aqueous solution, distilled water, and diethylether and then dried in a desiccator with silica gel. The dried microspheres were used as MS-A. Further, reacetylated microspheres (MS-B), double-layered microspheres (MS-C), and chitin powder-coated microspheres (MS-D) could be prepared by the modification of MS-A as described in the following. MS-B: MS-A (500 mg) was immersed in acetic anhydride of 1.5 ml for 45 minutes at room temperature, and then separated, washed successively with ethanol and ether, and then dried; MS-C: MS-B (250 mg) was dispersed in 1.5% (w/v) chitosan dissolved in 2% acetic acid aqueous solution of 50 ml, and then the preparation method of MS-A was repeated; MS-D: After the chitosan/acetic acid aqueous solution containing 5-FU was dropped into rapeseed oil, chitin powder (0.75 mg) was added. All other processes were the same as for MS-A.

Figure 1 shows the scanning electron micrographs of the various types of chitosan microspheres (MS-A, B, C, and D). Further, they can also be illustrated in Figure 2. The drug content was determined by

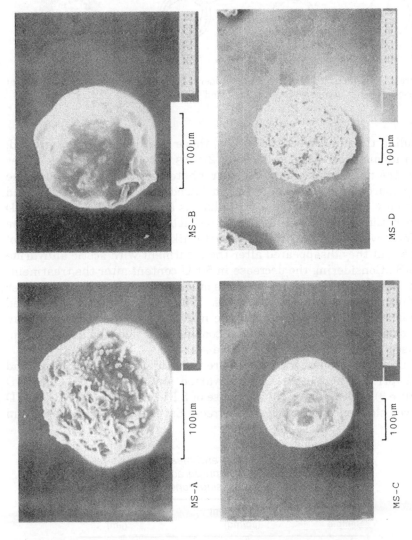

Figure 1. SEM photomicrographs of the different samples of chitosan microspheres. Key: see Table 1.

207

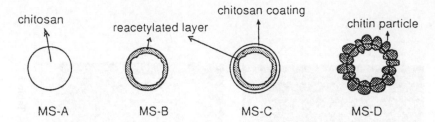

Figure 2. Hypothetical structures of microspheres drawn for an explanation of different release properties of microspheres A, B, C, and D.

grinding the microspheres. Namely, these microspheres were ground, put into distilled water, and left for 2 days at room temperature. After that, the filtrate was checked spectrophotometrically at 265 nm. The mean particle size was determined based on 200 particles selected randomly from the photomicrographs. As a result, MS-A, B, C, and D showed the drug content and mean particle size as described in Table 1. There were observed some fine crystal-like particles on the surface of MS-A, but they disappeared after the treatment with acetic anhydride (MS-B). Considering the decrease in 5-FU content after the treatment, they were thought to be fine crystals of 5-FU. MS-C showed a smoother surface than other microspheres. The drug content decreased remarkably as the treatment of the microspheres increased. Further, the operation of double chitosan coating and addition of chitin increased markedly the mean particle size.

Drug release behaviors, which were investigated using JP XI second fluid (pH 6.8) at $37 \pm 1°C$, were varied among MS-A, B, C, and D (Figure 3). MS-A showed rapid release of 5-FU, whereas MS-B, C, and D showed slow release to varying degrees. Especially, drug release from

Table 1. 5-FU contents and mean particle size of microspheres prepared by the different processing conditions.

Sample	5 FU Content (μg/mg)	Green Diameter (μm)
A	17.14	173.0
B	4.05	167.6
C	2.79	283.4
D	8.70	380.4

Samples: A, ordinary microspheres; B, reacetylated microspheres; C, double-layered microspheres; D, chitin powder-coated microspheres.

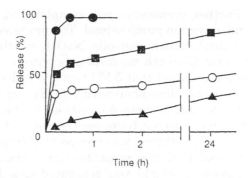

Figure 3. Release profiles of 5-FU from chitosan microspheres (sized by sieving in 150 to 420 μm) prepared by the different processing conditions in the JP XI second fluid (pH 6.8) at 37°C. ●, A; ■, B; ▲, C; and ○, D.

MS-C and D was considerably suppressed. These results indicated that the release characteristics of the microspheres could be modified through the treatments carried out on MS-B, C, and D.

Biodegradation of Chitosan Microspheres

The dry-in-oil method was applied to the preparation of the chitosan microspheres as follows: In this preparation, silicone oil, Dow Corning 360 Medical Fluid (R) (Dow Corning Co., Ltd.) was used as an oil phase. The specific gravity of this oil was slightly smaller than water, whereas other oils often used showed a relatively smaller density than water (Table 2). Probably because of this near density of silicone oil to water, the aqueous droplet containing chitosan was well dispersed in the oil, and microspheres could be obtained easily. 5-FU was dissolved into 5% chitosan (56% deacetylated) solution/1% acetic acid aqueous solution at the ratio of 20% (w/w) to chitosan. Twenty-five milliliters of the chitosan solution was dropped into 500 ml of silicone oil and then stirred at 500 rpm under reduced pressure by an aspirator at room tempera-

Table 2. Specific gravity of disperse media.

Disperse Medium	Specific Gravity d_{25}^{25}
Rapeseed oil	0.906–0.920
Liquid paraffin	0.860–0.890
Silicone oil	0.962–0.972

ture overnight. Further, evaporation was completed at 50°C using an aspirator first and a vacuum pump second. The produced microspheres were washed by diethylether, aqueous NaOH solution, water, and ethanol in that order. Chitosan microspheres with no drug were prepared in the same way except that 5-FU was not added.

Fifty milligrams of the microspheres with no drug were placed into a cylindrical filter paper and immersed in 100 ml of phosphate buffered saline of pH 7.4 with or without lysozyme chloride (1.5×10^{-2} mg/ml) at 37°C. When the outer solution was stirred at 200 rpm at 37°C, the microspheres were collected at appropriate times, dried in vacuo, and weighed. The decrease in weight was estimated as a degradation extent. The degradation profiles are described in Figure 4. The weight decrease was observed to a small extent even without lysozyme, which was supposed to be due to the dissolution of the water-soluble chitosan of a low molecular weight. Obviously, the degradation was observed to be accelerated by lysozyme. The microspheres with a smaller particle size underwent a more extensive degradation by lysozyme. This was considered to be because the surface area per weight was larger in the microspheres with a smaller particle size.

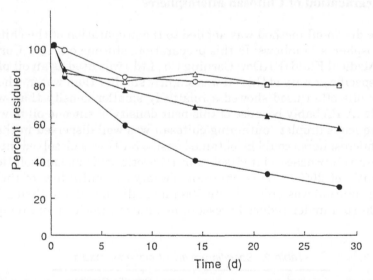

Figure 4. In vitro degradation profiles of the microspheres with different particle size in the incubation in phosphate buffered saline of pH 7.4 at 37°C with and without lysozyme. The microspheres with the mean diameter of 173 μm and 785 μm were obtained under the condition using silicone oil of 200 cs and 100 cs, respectively. ●: 173 μm, with lysozyme; ○: 173 μm, without lysozyme; ▲: 785 μm, with lysozyme; and △: 785 μm, without lysozyme.

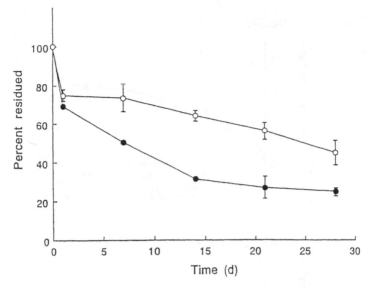

Figure 5. Degradation profiles of the microspheres with and without FU after subcutaneous implantation into the back of mice ($n = 3$). The results are expressed as the mean ± S.E., and S.E. is described when larger than the plotted point. ●: microspheres with FU and ○: microspheres without FU.

The in vivo degradation of microspheres was checked using the microspheres with 9.3% 5-FU or no drug, the mean particle sizes of which were 515 μm and 325 μm, respectively. Microspheres, 50 mg, were suspended in a small amount of olive oil and subcutaneously implanted into the back of a male BDF$_1$ mouse. The mouse was sacrificed at an appropriate time and all the remaining microspheres were taken out and then washed with purified water, acetone, and chloroform in that order. After drying in vacuo, their weight was measured.

The microspheres degraded gradually after subcutaneous implantation (Figure 5), which suggested biodegradation. The drug-loaded microspheres degraded faster than those with no drug. This might be due to the increase in the surface area following the release of 5-FU. The result demonstrated that chitosan microspheres could degrade gradually over 1 month.

These results indicate that biodegradation of chitosan microspheres is quite slow, and therefore, the control of release behavior is very important. The above system in which a drug is dispersed in the matrix of chitosan will be useful to obtain the chitosan microspheres with sustained-release properties. Microspheres composed of chitosan conjugated with a drug have also been prepared and evaluated. The release

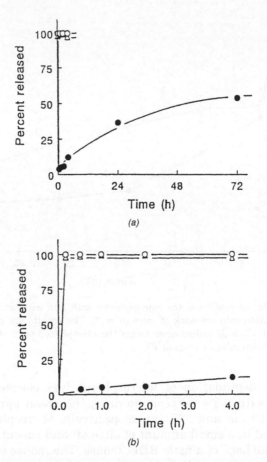

Figure 6. Drug release profiles from chi/FUR-m during the incubations in PBS of pH 7.4 (△) and 1/15 M phosphate buffer of pH 7.4 (○) at 37°C and from chi-glu-FUR-m during the incubation in 1/15 M phosphate buffer of pH 7.4 (●) at 37°C. (a) Release in the incubation for 0–72 H. (b) Release in the incubation for 0–4 h.

behavior of such beads is controlled mainly by the cleavage rate of the chemical bond between a drug and polymer support [22]. In this case, gradual drug release could be attained (Figure 6). This approach is considered to be very useful for the development of chitosan microspheres.

CHITOSAN-DRUG CONJUGATES

Preparation and Evaluation of Chitosan-Cytarabine Conjugate

Since chitosan is soluble in dilute aqueous acetic acid solution and possesses many reactive amino groups, it might be utilized as a polymer support for a macromolecular prodrug. Cytarabine (ara-C) is rapidly inactivated by cytidine deaminase and excreted quickly from the body [23,24]; therefore, to optimize its efficacy, a complex schedule of frequent administration or continuous prolonged infusion is required. Chemical modification of ara-C with chitosan is considered to be useful for overcoming its disadvantage. Conjugate of ara-C with chitosan was prepared as follows [25]: First, N^4-(4-carboxybutyryl)-ara-C (glu-ara-C) was synthesized. 5'-O-trityl-ara-C (2.06 mmol) and glutaric anhydride (3.09 mmol) were mixed in dioxane (48 ml) and stirred at room temperature for 1 week, and then the solvent was evaporated. Detritylation was carried out by the treatment of the residue by 50% (v/v) acetic acid aqueous solution (30 ml) at 90°C for 5–10 minutes. After separation of the precipitate, the filtrate was evaporated to dryness and then chromatographed with a silica gel column using a mixture of chloroform, methanol, and acetic acid (16:4:1, v/v) as an elution solvent. The fractions containing new compound were collected and then the solvent was evaporated. Glu-ara-C was obtained by recrystallization of the residue from ethanol. Glu-ara-C could also be synthesized by the direct reaction of ara-C (4.09 mmol) with glutaric anhydride (4.39 mmol) in anhydrous pyridine (30 ml) at room temperature for 2 days. After evaporation of the solvent, the residue was separated by a silica gel column using a mixture of chloroform, methanol, and acetic acid (125:20:4, v/v) as an elution solvent. The obtained product was recrystallized from ethanol to give glu-ara-C. Second, conjugate of glu-ara-C with chitosan (chi-glu-ara-C) was prepared by carbodiimide coupling in aqueous solution. Chitosan (MW $8.0 \times 10^5 - 1.5 \times 10^6$) was dissolved in HCl aqueous solution of pH 3 and then adjusted to pH 6 with 0.2% NaOH aqueous solution. 1-Ethyl-3-(3-dimethylaminopropyl)carbodiimide hydrochloride (EDC) and glu-ara-C were added to the chitosan solution at pH 6.0 and stirred at room temperature for 24 h. After that, small molecules were ultrafiltered using an ultrafilter membrane (MW 50,000). The residue was washed with 0.1 M NaCl and purified water, and then the mixture pH was adjusted to 7.2–7.3 using 0.2% NaOH aqueous solution. Addition of acetone to the mixture gave the precipitate, and it was dried in vacuo to obtain the product (chi-glu-ara-C). The content of ara-C in the conjugate was investigated as follows: Five milligrams of the conjugate was added in 5 ml of 1 N NaOH aqueous solution and heated in a water bath at 40°C for 10 minutes. Since this operation caused complete regeneration of ara-C, the ara-C content could be estimated from the UV absorption of the filtrate after the heat operation. Structures of ara-C, glu-ara-C, and chi-glu-ara-C are shown in Figure 7. The relationship between the reaction condition and drug content is described in Table 3.

The drug release profiles of chi-glu-ara-C are shown in Figure 8. After

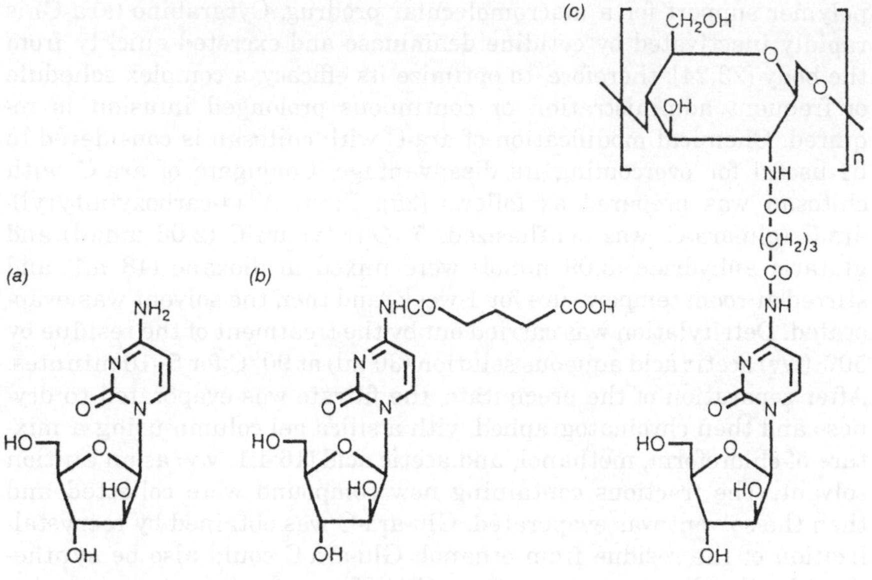

Figure 7. Structures of (a) ara-C, (b) glu-ara-C, and (c) chi-glu-ara-C.

incubation for 7 days at 37°C, the amounts of ara-C released from the conjugate were 21.7% at pH 6, 56.0% at pH 7.4, and 76.2% at pH 8. 1-β-D-Arabinofuranosyluracil (ara-U) and glu-ara-C were generated slightly. Since the release rate of ara-C from glu-ara-C was very slow at pH 7.4 [25], most of the regenerated ara-C was considered to be released directly from chi-glu-ara-C. Thus, chi-glu-ara-C showed good drug release under the physiological condition. It could be suggested that chi-glu-ara-C should be an available macromolecular prodrug of ara-C for gradual drug release.

Acute toxicities of chitosan and chi-glu-ara-C (4.9% drug content) were checked based on survivors on the thirty-fifth day after the single

Table 3. Preparation conditions and the ara-C contents of the conjugates.[a]

Glu-ara-C (mg)	Chitosan (mg)	EDC (g)	Purified Water (ml)	Content of ara-C (%, w/w)
49.8	50.8	1.0013	20	4.9
555.9	551.0	10.0104	140	5.0
594.2	594.1	11.9500	300	3.3
558.2	298.2	6.0164	150	5.6

[a]Every reaction was carried out at room temperature at pH 6.0 for 24 h.

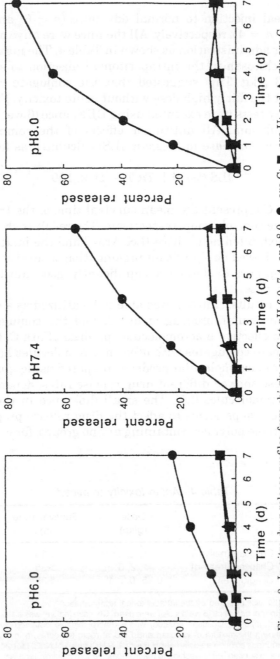

Figure 8. In vitro drug release profiles from chi-glu-ara-C at pH 6.0, 7.4, and 8.0. ●, ara-C; ■, ara-U; and ▲, glu-ara-C.

215

intraperitoneal injection to normal ddy mice ($n = 6$) and a normal BDF_1 mouse ($n = 1$), respectively. All the mice were alive on the thirty-fifth day after administration as shown in Table 4. The maximum tolerable dose of chitosan in the intraperitoneal injection to a mouse was more than 3 g/kg. This suggested that the conjugate could be administered at the fairly high dose without acute toxicity. Further, antitumor activity tests were executed using BDF_1 mice (6 weeks old) bearing P388 leukemia. All antitumor effects of the compounds were estimated by an increase in lifespan (ILS) calculated as follows:

$$ILS\ (\%) = [(T/C) - 1] \times 100$$

where T and C represent the mean survival time of the treated group and that of the control group, respectively. The result is shown in Table 5. Chitosan exhibited no positive ILS. Ara-C and the mixture of ara-C and chitosan showed no significant prolongation of survival time to the control. Chi-glu-ara-C showed a significantly good antitumor effect even at a low dose.

From these results, chitosan was able to be utilized as a drug carrier for a macromolecular prodrug of ara-C, and the conjugate, chi-glu-ara-C, was an effective macromolecular prodrug of ara-C. Glu-ara-C is useful to prepare conjugates with other macromolecules having amino groups. Other macromolecular prodrugs prepared using glu-ara-C were reported. These showed different drug release rates, depending on the kinds of macromolecules [25]. The small difference in the structures around amino groups seems to affect the drug release properties. Chitosan is a suitable polymer, containing amino groups, for conjugation of ara-C.

Table 4. Acute toxicity to mice.[a]

	Dose (g/kg)	Survival Time (d)
Chitosan (native)[b]	1, 2, 3	>35
Chitosan (treated)[b]	1, 2, 3	>35
Chi-glu-ara-C[c]	0.25, 0.5, 1[d]	>35

[a]This was checked by the intraperitoneal administration.
[b]Normal ddY mice (6 weeks old) were used. Chitosan (native) was an intact chitosan. Chitosan (treated) was a macromolecule obtained by stirring the mixture of chitosan and EDC at room temperature in an aqueous solution adjusted at pH 6 with 1 N HCl and 1 N NaOH for 24 h and subsequently washing to macromolecule with an ultrafilter membrane, UK-50 used for the preparation of the conjugate.
[c]A normal BDF_1 mouse (6 weeks old) was used for each dose.
[d]This dose corresponded to 44 mg eq ara-C/kg.

Table 5. *Effect of chi-glu-ara-C and ara-C on the survival time of mice bearing P388 leukemia.*[a]

Material	Dose (mg eq ara-C/kg)	Survival Days (mean ± S.D.)	ILS (%)	Control[b]
Ara-C	50	11.2 ± 1.2	9.8	A
	100	12.0 ± 1.2	3.4	C
Chitosan[c]	—	8.4 ± 0.5	−20.8	B
	—	9.2 ± 0.8	−20.7	C
Ara-C + chitosan[c]	50	13.4 ± 3.6	26.4	B
	100	11.6 ± 2.1	0.0	C
Chi-glu-ara-C	44	15.2 ± 2.6[d]	43.4	B
	88	22.5 ± 3.5[d]	60.7	D

[a]P388 leukemia cells were inoculated intraperitoneally into BDF₁ mice. Chemotherapy was carried out with a single intraperitoneal injection at 24 h after inoculation.
[b]The survival days (mean ± S.D.) of the control groups: A (n = 6), 10.2 ± 1.7; B (n = 5), 10.6 ± 1.3; C (n = 5), 11.6 ± 1.5; D (n = 4), 14.0 ± 1.2. The control groups received an intraperitoneal injection of 0.5–1.0 ml of normal saline.
[c]Chitosan was administered at an equivalent dose of chi-glu-ara-C by a single intraperitoneal injection.
[d]$p < 0.01$ versus the control group.

Drug Release and Antitumor Characteristics of *N*-Succinyl-Chitosan-Mitomycin C Conjugate and 6-*O*-Carboxymethyl-Chitin-Mitomycin C Conjugate

N-Succinyl-chitosan (Suc-chitosan) and 6-*O*-carboxymethyl-chitin (CM-chitin) are water-soluble chitosan and chitin derivatives, respectively. These polymers are of low toxicity due to their structures, and they are useful as macromolecules for conjugation with drugs due to the presence of reactive carboxyl groups.

Mitomycin C (MMC) is widely used in cancer chemotherapy, but it presents such side effects as severe bone marrow depression and gastrointestinal damage [26]. These disadvantages can be overcome by means of concentrating the cytotoxicity at the tumor site and achieving prolonged duration of the activity there. Therapeutic efficacy will be enhanced when the burden on other tissues is minimized by improving the pharmacokinetic and pharmacodynamic properties of the drug. The conjugation of MMC with Suc-chitosan and CM-chitin is proposed as a possible approach to realize the above improvement.

Suc-chitosan-MMC conjugate (Suc-chitosan-MMC) and CM-chitin-MMC conjugate (CM-chitin-MMC) were prepared as follows: MMC (60 mg), Suc-chitosan (480 mg), and EDC (1.2 g) were mixed in water (150 ml), the pH of which was adjusted to 5 using 1% HCl aqueous solution and stirred at room temperature for 45 minutes. After that, the resulting precipitate was isolated by filtration and washed with water. Suc-chitosan-MMC was obtained by drying the residue. MMC (60 mg), CM-chitin (480 mg), and EDC (1.2 g) were mixed in water (150 ml), the pH of which was adjusted to 5 using 1% HCl aqueous solution and stirred at room temperature for 2 h. The product was precipitated by addition of acetone. The precipitate was washed by a mixture of acetone and water (3:1, v/v). CM-chitin-MMC was obtained by drying the residue. The drug content of the conjugates was checked as follows: The aliquot of the mixture was withdrawn at the end of conjugation reaction and then diluted by addition of 3 ml of water. After the diluted sample was ultrafiltered (MW 50,000), the amount of MMC in the filtrate was determined spectrophotometrically at 364 nm. Thus, the amount of unbound MMC was determined, and then the binding ratio of MMC to Suc-chitosan or CM-chitin was calculated. As a result, Suc-chitosan-MMC (drug content, 12%) and CM-chitin-MMC (drug content, 11%) were obtained. Figure 9 shows the structures of Suc-chitosan-MMC and CM-chitin-MMC. Suc-chitosan-MMC was water insoluble, and CM-chitin-MMC was partially water soluble. The remaining amino groups and carboxyl groups in Suc-chitosan were considered to be combined together by carbodiimide coupling and consequently to form cross-linking, whereas for CM-chitin, such cross-linking would occur to a small extent because of few amino groups.

In vitro drug release from the conjugates was checked in 1/15 M phosphate buffer of pH 7.4 (Figure 10). Each conjugate showed mono-exponential liberation of MMC. The apparent release rate constants (k_{app}) were remarkably different between Suc-chitosan-MMC and CM-chitin-MMC (Table 6). Conversion reaction was analyzed by the linear kinetics as described in Chart 1, and the release rate (k_2) was found similar to k_{app}, which was because the degradation of MMC was slight under the experimental condition. Thus, Suc-chitosan-MMC can be characterized as a conjugate, being water insoluble but swelling in aqueous solution and gradually regenerating MMC at physiological pH. On the other hand, CM-chitin-MMC was partially water soluble and exhibited a relatively fast drug release.

The in vivo release was investigated as one of the important bio-

Table 6. Apparent release rate constants (k_{app}s) and apparent half-lives [$t(1/2)_{app}$s] in 1/15 M phosphate buffer, pH 7.4, at 37°C.

Conjugate	k_{app} (h^{-1})	$t(1/2)_{app}$ (h)
Suc-chitosan-MMC	3.9×10^{-3}	180
CM-chitin-MMC	1.1×10^{-1}	6.2

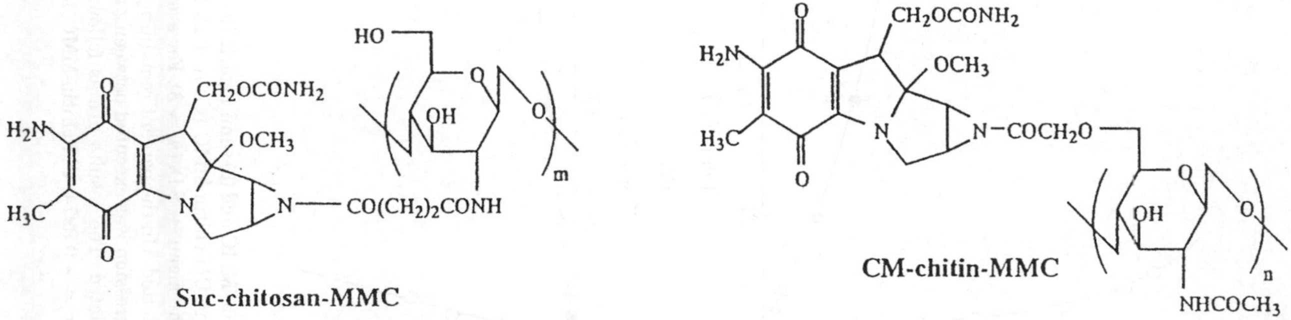

Suc-chitosan-MMC

CM-chitin-MMC

Figure 9. Chemical structures of Suc-chitosan-MMC and CM-chitin-MMC.

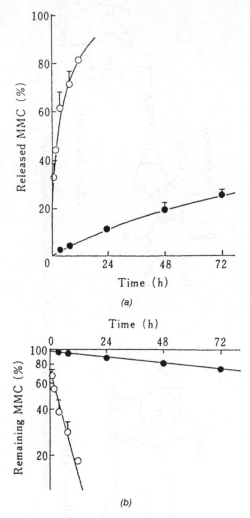

(a)

(b)

Figure 10. Release profiles of (a) MMC and (b) semi-logarithmic plots of remaining MMC for Suc-chitosan-MMC (●) and CM-chitin-MMC (○) in 1/15 M phosphate buffer of pH 7.4. Every point represents the mean ± S.D. ($n = 3$). For every point, S.D. is described when larger than the point plotted. On the percent remaining MMC (x) and the incubation time, h, (t), the linear regression was executed between t and log x, and the correlation coefficient (r) was calculated. The results are as follows. Suc-chitosan-MMC, log $x = 1.992 - 0.001712t$, and $r = -0.9974$; CM-chitin-MMC, log $x = 1.838 - 0.04839t$, and $r = -0.9879$.

Chart 1. Conversion Reaction Scheme for Conjugates

$$(MMC) = (k_2 A_0/(k_1 - k_2 - k_3)) \exp(-(k_2 + k_3)t)$$

$$+ (100 - (k_1 - k_3)A_0/(k_1 - k_2 - k_3)) \exp(-k_1 t)$$

in which (MMC) means percent of free MMC in the incubation medium to the initial contained MMC and A_0 means 100 minus percent of burst MMC.

pharmaceutical properties. Male Wistar rats (200–250 g) were anesthetized by urethane and fixed on their backs. The suspension of Suc-chitosan-MMC or CM-chitin-MMC (15 mg MMC eq/kg) was administered intraperitoneally at 1.5 ml/100 g body weight. The solution of MMC (5 mg/kg) was injected intraperitoneally at 0.4 ml/100 g body weight. Each blood sample (0.5 ml) was withdrawn from the jugular vein at appropriate times, and the plasma was obtained after centrifugation. MMC in the plasma was extracted with the addition of 2 ml of the mixture of chloroform and 2-propanol (1:1, w/w) to 0.2 ml of the plasma. The clear supernatant was evaporated to dryness at 40°C under nitrogen gas. After dissolving the residue in 100 μl methanol, 20 μl of the aliquot was injected on reversed phase HPLC. Figure 11 shows the plasma concentration-time courses of MMC. MMC exhibited a maximum concentration of 1.45 μg/ml at 30 minutes after injection, and then the concentration of MMC decreased rapidly. CM-chitin-MMC showed a maximum concentration of MMC at 2 h after injection, and the MMC level decreased more slowly. Concerning Suc-chitosan-MMC, the maximum plasma concentration of MMC was much lower than that for MMC and CM-chitin-MMC, but the plasma concentration of MMC was maintained at almost a constant level over 24 h after injection. These results suggested that MMC should enter the blood circulation quickly, CM-chitin-MMC released MMC gradually but relatively fast, and Suc-chitosan-MMC released MMC very slowly. The pharmacokinetic analysis using depot models supported these suggestions [19]. The in vivo release rates from CM-chitin-MMC and Suc-chitosan-MMC were consistent with the in vitro drug release rates. Each conjugate showed good retention in the body, but some part seemed to be eliminated or degraded before release of MMC, which brought about the decrease in AUC.

The antitumor activity tests were performed using BDF$_1$ mice (6 weeks old) bearing L1210 leukemia. At 24 h after L1210 leukemia cells (1 × 10^5) were inoculated intraperitoneally per mouse, the conjugates and MMC were administered intraperitoneally. As a control, saline was administered. Antitumor activity was evaluated by comparing the

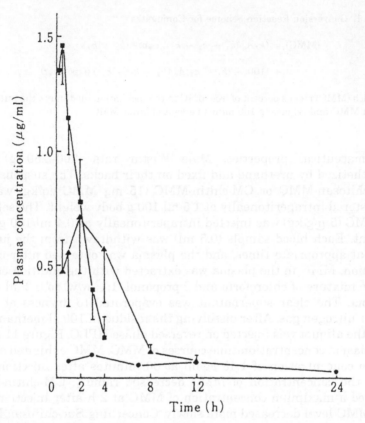

Figure 11. Plasma concentration-time profiles of MMC after single intraperitoneal bolus injections of MMC (5 mg/kg), Suc-chitosan-MMC (15 mg eq MMC/kg) and CM-chitin-MMC (15 mg eq MMC/kg). Each point represents the mean ± S.E. (n = 4). S.E. is shown unless smaller than the point plotted. ■, MMC (5 mg/kg); ●, Suc-chitosan-MMC (15 mg eq MMC/kg); ▲, CM-chitin-MMC (15 mg eq MMC/kg). A maximum blood concentration of MMC (C_{max}) for MMC was significantly different from that for Suc-chitosan-MMC ($p < 0.01$). C_{max} for MMC was significantly different from that for CM-chitin-MMC ($p < 0.05$).

mean survival time of the treated mice (T) with that of the control mice (C), i.e., by calculating the increase in lifespan (ILS), $[(T/C) - 1] \times 100$ (%). Table 7 summarizes the antitumor activities of Suc-chitosan-MMC and CM-chitin-MMC against L1210 leukemia in comparison with that of MMC. At the low dose, each conjugate showed lower effects than MMC. At the high dose, each conjugate became more effective, but the effect of MMC decreased. The antitumor effect seemed to correlate with the drug release rate. That is to say, quick and extensive exposure of MMC by administration of free MMC gave a high-antitumor effect at a low dose but side effects at a high dose. A prolonged and low supply of MMC by administration of the conjugate, especially Suc-chitosan-MMC, showed a better effect at a high dose.

Biopharmaceutical Properties of *N*-Succinyl-Chitosan-Mitomycin C Conjugate and 6-*O*-Carboxymethyl-Chitin-Mitomycin C Conjugate as Microparticles

N-Succinyl-chitosan-mitomycin C conjugate (Suc-chitosan-MMC) and 6-*O*-carboxymethyl-chitin-mitomycin C conjugate (CM-chitin-MMC) were obtained as water-insoluble and partially water-soluble products. Therefore, they were suggested to be microparticulated by being ground. The particulate carrier system is important as a drug delivery system for such a targeting as chemoembolization [27,28]. The characteristics of the particle size of the conjugates and their in vivo distribution were investigated after injection into a vein.

Suc-chitosan-MMC and CM-chitin-MMC were prepared by the same method, and their MMC contents were analyzed to be 9.8% and 9.7%, respectively. Suc-chitosan-MMC and CM-chitin-MMC (5 mg) were ground to fine powder in 10 ml of normal saline using a glass homogenizer. The shape and size of the conjugate particles contained in the suspension were observed with an optical microscope. The sizes (Heywood diameter) were measured randomly for about 500 particles,

Table 7. Effect of the conjugates and MMC on the survival time of mice bearing L1210 mice.[a]

Material	Dose (mg eq MMC/kg)	Survival Days (mean ± S.D.)	ILS (%)	Survivors at 30 d
Control		7.8 ± 0.5		0/6
MMC	2.5	16.2 ± 7.9[b]	107.7	1/6
	5.0	14.4 ± 1.3[d]	84.6	0/6
	10.0	12.2 ± 3.9[b]	54.6	0/6
Suc-chitosan-MMC	5.0	11.3 ± 2.4[b]	45.3	0/6
	10.0	11.2 ± 0.4[d]	43.2	0/6
	20.0	12.8 ± 1.2[b]	64.5	0/6
CM-chitin-MMC	2.5	10.0 ± 1.1[c]	28.1	0/6
	5.0	11.3 ± 2.3[b]	45.3	0/6
	10.0	13.3 ± 1.0[d]	70.9	0/6

[a]Chemotherapy was carried out by a single intraperitoneal administration at 24 h after the intraperitoneal tumor inoculation.
[b]$p < 0.05$ versus the control group.
[c]$p < 0.01$ versus the control group.
[d]$p < 0.001$ versus the control group.

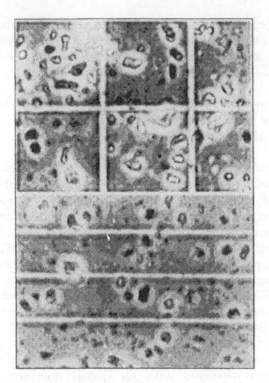

Figure 12. The optical photomicrograph of carboxymethyl-chitin-mitomycin C (up) and *N*-succinyl-chitosan-mitomycin C (down).

and the arithmetic average of the particle diameter was calculated by $d_{av} = \Sigma(n \times d)/\Sigma n$. Suc-chitosan-MMC and CM-chitin-MMC were obtained as purple powder after the homogenization. Although both conjugates showed irregular shape, their shape was not very different from that of a cube or sphere (Figure 12). The size distributions were obtained as illustrated in Figure 13. The averaged diameters of Suc-chitosan-MMC and CM-chitin-MMC were 4.1 μm and 7.1 μm, respectively.

The obtained suspensions (0.5 ml) containing Suc-chitosan-MMC and CM-chitin-MMC (5 mg) were injected intravenously via tail vein into ddy male mice (6 weeks old). At the appropriate time periods after injection, the mice were sacrificed, and the lungs, liver, kidneys, and blood were taken. The organs were weighed, and after addition of isotonic phosphate buffer of pH 9.0 with twofold weight of their organs, they were homogenized and incubated for 1 h at 90°C for Suc-chitosan-MMC and for 3 h at 37°C for CM-chitin-MMC to completely release

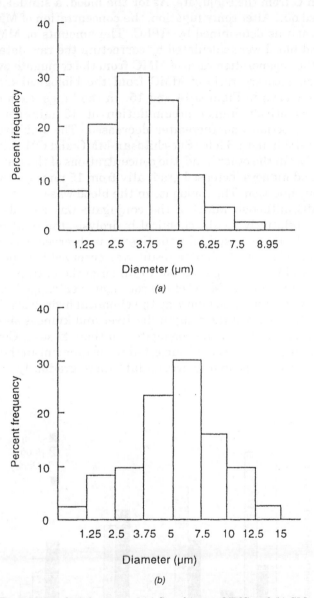

(a)

(b)

Figure 13. Particle size distributions of (a) Suc-chitosan-MMC and (b) CM-chitin-MMC.

225

mitomycin C from the conjugate. As for the blood, a similar operation was carried out. After centrifugation, the concentration of MMC in the supernatant was determined by HPLC. The amounts of MMC in the tissues and blood were calculated by correcting the raw data of MMC based on the regeneration ratio of MMC from the conjugate by hydrolysis and the recovery ratio of MMC from the biological media. The results are shown in Figures 14 and 15. In the lungs, each conjugate showed a relatively higher accumulation at 15 minutes after intravenous injection and thereafter decreased. The half-lives in the lungs were 4.3 h and 1.6 h for Suc-chitosan-MMC and CM-chitin-MMC, respectively. On the other hand, the concentrations of the conjugates in the liver and kidneys increased gradually from 15 minutes to 5 h after intravenous injection. The conjugate in the blood was slight. As to CM-chitin-MMC, although MMC of the conjugate was considered to be eliminated relatively fast by chemical hydrolysis, the total remaining amount in the tissues, especially the liver, was observed to be high even at 5 h after injection. A similar result was observed for Suc-chitosan-MMC. Free MMC was reported not to undergo the entrapment in the tissues [29]. In addition, MMC of the conjugates entrapped in the liver or kidneys was eliminated following the chemical hydrolysis. Therefore, the gradual increase of the drug in the liver and kidneys signified the gradual accumulation of the conjugates to their tissues. Considering that the smaller particles were subjected to entrapment in the liver, the smaller particles appeared to accumulate there gradually.

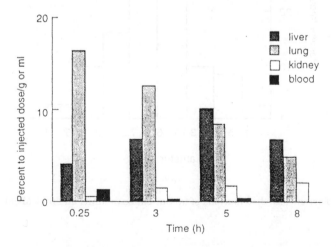

Figure 14. Tissue distribution of N-succinyl-chitosan-mitomycin C at 15 min, 3, 5, and 8 h after intravenous injection via tail vein of mice. All data represent the mean ($n = 3$).

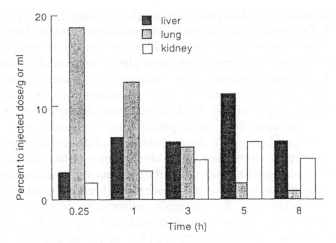

Figure 15. Tissue distribution of carboxymethyl-chitin-mitomycin C at 15 min, 3, 5, and 8 h after intravenous injection via tail vein of mice. All data represent the mean ($n = 3$).

Generally, it is known that larger microspheres (7–15 μm) are distributed mainly in the lungs and smaller microspheres (0.3–3 μm) mainly in the liver, and the distribution in the blood is very slight [9,30,31]. Suc-chitosan-MMC and CM-chitin-MMC showed relatively high accumulation in the lungs in the initial stage, and the accumulation of the conjugate in the liver increased gradually in the latter stage. Since both conjugate particles showed irregular shapes and swelling properties, their behaviors in the body distribution might not be the same as that of simple microspheres. Since the total amount distributed in the tested tissues was relatively small, especially in the initial stage, their accumulation in the other tissues, i.e., heart, spleen, and blood vessel was considered to be feasible. These points will be made clear by a check of the accumulation in the other tissues. The pathological influence of the conjugate suspensions on the lungs, liver, and kidneys was confirmed to be of low toxicity from the histological study [20]. These results suggest that the water-insoluble conjugates might be controlled as to disposition in the body by changing the size of the conjugate particles. These properties will be useful in the chemoembolization of the ill site.

FUTURE ASPECTS

Chitosan microspheres will have to be more controllable with regards to drug release. For this purpose, a mixture or coating of chitosan with other suitable polymers might be useful, or microspheres composed of or containing chitosan-drug conjugates might be adequate. Further, since chitosan microspheres are known to be swellable, the body fluid will permeate into them, and then complete obturation may be evitable. If these characteristics diminish the obstacles presented by microspheres, the administration of the microspheres into the vascular systems might be permitted. Further examination of the particle size distribution, accumulation, and pathological influence will help to clarify these problems. Chitosan, chitin, or chitosan derivative-drug conjugates may be developed as to therapeutic antitumor agents or strong bioactive agents. Since the disposition or localization in the body is dependent on the physicochemical characteristics of the polymer, its properties, for example, adsorption of chitosan to the anionic components of the tissues or cells [32,33], can be utilized for more effective use. The conjugate will be useful as a molecular drug delivery system for specific targeting. The accumulative and immunogenic properties are serious problems for practical use. Further precise and detailed construction, characterization, and comprehension of biological responses will raise the utility of these conjugates.

ACKNOWLEDGEMENTS

The authors are very grateful to Mrs. Y. P. Li, Mr. T. Yoshino, Mr. H. Ichikawa, and Dr. Y. Song for their cooperation in this work.

REFERENCES

1. Prudden, J. F., Miegel, P., Hanson, P., Friedlich, L. and Balassa, L. 1970. "Discovery of a Potent Pure Chemical Wound-healing Accelerator," *Am. J. Surg.,* 119(5):560–564.

2. Davies, R. C., Neuberger, A. and Wilson, B. M. 1969. "The Dependence of Lysozyme Activity on pH and Ionic Strength," *Biochim. Biophys. Acta,* 178:294–305.

3. Amano, K. and Ito, E. 1978. "The Action of Lysozyme on Partially Deacetylated Chitin," *Eur. J. Biochem.,* 85(1):97–104.

4. Pangburn, S. H., Trescony, P. V. and Heller, J. 1982. "Lysozyme Degradation of Partially Deacetylated Chitin, Its Films and Hydrogels," *Biomaterials,* 3(2):105–108.

5. Nakamura, F., Onishi, H., Machida, Y. and Nagai, T. 1992. "Lysozyme-catalyzed Degradation Profiles of the Conjugates between Chitosans Having Some Deacetylation Degrees and Methotrexate," *Yakuzaigaku,* 52(1):59–67.

6. Ochi, A., Shibata, S., Mori, K., Sato, T. and Sunamoto, J. 1990. "Targeting

Chemotherapy of Brain Tumor Using Liposome-encapsulated Cisplatin–Part 2. Pullulan-coated Liposomes to Target Brain Tumor," *Drug Deliv. Syst.*, 5(4):261–265.

7. Ueno, M., Zou, Y. Y., Ono, A., and Horikoshi, I. 1991. "Basic Study on Hepatic Artery Chemoembolization Using Temperature-sensitive Liposome in the Treatment of Hepatic Tumor: Animal Experiment," *Jpn. J. Cancer Chemother.*, 18(12):2183–2185.

8. Ono, A., Ueno, M. and Horikoshi, I. 1992. "Basic Study on Hepatic Artery Chemoembolization Using Temperature-sensitive Liposome for Treatment of Hepatic Tumor–Formation of Embolus with Liposome in Hepatic Artery," *Drug Deliv. Syst.*, 7(1):31–36.

9. Kanke, M., Simmous, G. H., Weiss, D. L., Bivins, B. A. and Deluca, P. P. 1980. "Clearance of ^{141}Ce-labeled Microspheres from Blood and Distribution in Specific Organs Following Intravenous and Intraarterial Administration in Beagle Dogs," *J. Pharm. Sci.*, 69(7):755–762.

10. Widder, K. J., Senyei, A. E. and Ranny, D. F. 1979. "Magnetically Responsive Microspheres and Other Carriers for Biophysical Targeting of Antitumor Agents," *Adv. Pharmacol. Chemother.*, 16:213–271.

11. Hara, T., Mikura, Y., Nagai, A., Miura, Y., Futo, T., Tomida, Y., Shimizu, H. and Toguchi, H. 1994. "Controlled Release of Thyrotropin Releasing Hormone from Microspheres: Evaluation of Release Profiles and Pharmacokinetics after Subcutaneous Administration," *J. Pharm. Sci.*, 83(6):798–801.

12. Ibrahim, M., Couvreur, P., Roland, M. and Speiser, P. 1983. "New Magnetic Drug Carrier," *J. Pharm. Pharmacol.*, 35(1):59–61.

13. Morimoto, Y., Okumura, M., Sugibayashi, K. and Kato, Y. 1981. "Biomedical Application of Magnetic Fluids. II. Preparation and Magnetic Guidance of Magnetic Albumin Microsphere for Site Specific Drug Delivery in Vivo," *J. Pharm. Dyn.*, 4(8):624–631.

14. Tanigawa, N., Satomura, K., Hikasa, Y., Hashida, M., Muranishi, S. and Sezaki, H. 1980. "Surgical Chemotherapy against Lymph Node Metastases," *Surgery*, 87(2):147–152.

15. Li, Y. P., Machida, Y., Sannan, T. and Nagai, T. 1991. "Preparation of Chitosan Microspheres Containing Fluorouracil Using the Dry-in-oil Method and Its Release Characteristics," *S. T. P. Pharma Sciences*, 1(6):363–368.

16. Yoshino, T., Machida, Y., Onishi, H. and Nagai, T. 1996. "Characteristics on the Drug Loading, Drug Release and Degradation of Chitosan Microspheres Prepared Using the Dry-in-oil Method," *S. T. P. Pharma Sciences*, 6(2):122–128.

17. Ichikawa, H., Onishi, H., Takahata, T., Machida, Y. and Nagai, T. 1993. "Evaluation of the Conjugate between N^4-(4-Carboxybutyryl)-1-β-D-arabinofuranosylcytosine and Chitosan as a Macromolecular Prodrug of 1-β-D-Arabinofuranosylcytosine," *Drug Des. Deliv.*, 10:343–353.

18. Song, Y., Onishi, H. and Nagai, T. 1992. "Synthesis and Drug-release Characteristics of the Conjugates of Mitomycin C with *N*-Succinyl-chitosan and Carboxymethyl-chitin," *Chem. Pharm. Bull.*, 40(10):2822–2825.

19. Song, Y., Onishi, H. and Nagai, T. 1993. "Pharmacokinetic Characteristics and Antitumor Activity of the *N*-Succinyl-chitosan-mitomycin C Conjugate and the Carboxymethyl-chitin-mitomycin C Conjugate," *Biol. Pharm. Bull.*, 16(1):48–54.

20. Song, Y., Onishi, H., Machida, Y. and Nagai, T. 1995. "Particle Characteristics of Carboxymethyl-chitin-mitomycin C Conjugate and *N*-Succinyl-chitosan-mitomycin C Conjugate and Their Distribution and Histological Effect on Some Tissues after Intravenous Administration," *S. T. P. Pharma Sciences*, 5(2):162–170.

21. Morishita, M., Inaba, Y., Fukushima, M., Hattori, Y., Kobari, S. and Matsuda, T. 1976. "Encapsulation of Medicaments," U.S. Pat. No. 3,960,757.

22. Onishi, H., Shimoda, J. and Machida, Y. 1996. "Chitosan-drug Conjugate Microspheres: Preparation and Drug Release Properties of Chitosan Microspheres Composed of Chitosan-2′ or 3′-(4-Carboxybutyryl)-5-fluorouridine (Chi-glu-FUR)," *Drug Dev. Ind. Pharm.*, 22(5):457–463.

23. Aoshima, M., Tsukagoshi, S., Sakurai, Y., Oh-ishi, J., Ishida, T. and Kobayashi, H. 1976. "N^4-Behenoyl-1-β-D-arabinofuranosylcytosine as a Potential New Antitumor Agent," *Cancer Res.*, 36(8):2726–2732.

24. Onishi, H., Seno, Y., Pithayanukul, P. and Nagai, T. 1991. "Antitumor Characteristics of the Conjugate of N^4-(4-Carboxybutyryl)-ara-C and Ethylenediamine-introduced Dextran and Its Resistance to Cytidine Deaminase," *Drug Des. Deliv.*, 6:272–280.

25. Pithayanukul, P., Onishi, H. and Nagai, T. 1989. "In vitro pH-dependent Release from N^4-(4-Carboxybutyryl)-1-β-D-arabinofuranosylcytosine and Its Conjugate with Poly-L-lysine or Decylenediamine-dextran T70," *Chem. Pharm. Bull.*, 37(6):1587–1590.

26. Carter, S. and Crooke, S., ed. 1980. *Mitomycin C Current Status and New Developments.* New York, Academic Press.

27. Milano, G., Boublil, J. L., Bruneton, J. N., Bourry, J. and Renee, N. 1985. "Systemic Blood Levels after Intra-arterial Administration of Microencapsulated Mitomycin C in Cancer Patients," *Eur. J. Drug Metab. Pharmacokinet.*, 10(3):197–301.

28. Sato, K., Moriyama, M., Kato, T., Goto, A. and Uno, K. 1991. "Monitoring of Systemic Blood Levels after Intra-arterial Administration of Microencapsulated Mitomycin C in Cancer Patients," *Drug Deliv. Syst.*, 6:191–194.

29. Takakura, Y., Takagi, A., Hashida, M. and Sezaki, H. 1987. "Disposition and Tumor Localization of Mitomycin C-dextran Conjugates in Mice," *Pharm. Res.*, 4(4):293–300.

30. Yoshioka, T., Hoshida, M., Muranishi, S. and Sezaki, H. 1981. "Specific Delivery of Mitomycin C to the Liver, Spleen and Lung: Nano- and Microspherical Carriers of Gelatin," *Int. J. Pharm.*, 81(2):131–141.

31. Sugibayashi, K., Morimoto, Y., Nadai, T., Kato, Y., Hasegawa, A. and Arita, T. 1979. "Drug-carrier Property of Albumin Microspheres in Chemotherapy. II. Preparation and Tissue Distribution in Mice of Microsphere-entrapped 5-Fluorouracil," *Chem. Pharm. Bull.,* 27(1):204–209.

32. Lehr, C.-M., Bouwstra, J. A., Schacht, E. H. and Junginger, H. E. 1992. "In vitro Evaluation of Mucoadhesive Properties of Chitosan and Some Other Natural Polymers," *Int. J. Pharm.,* 78(1):43–48.

33. Lueßen, H. L., Lehr, C. M., Retel, C.-O., Noach, A. B. J., De Boer, A. G., Verhoef, J. C. and Junginger, H. E. 1994. "Bioadhesive Polymers for the Peroral Delivery of Peptide Drugs," *J. Cont. Rel.,* 29(3):329–338.

Chitosan-Alginate Affinity Microcapsules for Isolation of Bovine Serum Albumin

Mattheus F. A. Goosen,[1] Osei-Wusu Achaw and
Edward W. Grandmaison
Department of Chemical Engineering
Queen's University
Kingston, Ontario, Canada K7L 3N6

INTRODUCTION

Product isolation from complex biological fluids and biotechnology-related processes are often not possible via conventional separation methods such as distillation, evaporation, or solvent extraction. This is due to two reasons. First, most of the target molecules occur in very low concentrations and are often adversely affected by the conditions utilized during conventional separation processes. Second, the target molecules must often be isolated from mixtures that contain other molecules of similar physical and chemical properties. These therefore require more selective methods than those attainable by conventional methods.

Affinity separation has been identified as capable of isolating these molecules to purity requirements of the marketplace [1–4]. In affinity separation the specific interaction between two molecules is employed for the separation of either of the pair from a mixture. Conventionally, one of the pair, the ligand, is immobilized on a solid support, which is

---

[1]Present address: Sultan Qaboos University, College of Agriculture, P.O. Box 34, AL-Khod 123, Sultanate of Oman.

then contacted with the mixture from which the target molecule must be isolated. Three distinct stages of affinity separation processes can be identified: adsorption, washing, and elution (desorption). In the adsorption stage the target molecule (adsorbate or ligate) selectively adsorbs onto the immobilized ligand. Subsequently, at the washing stage, the physically (nonspecifically) adsorbed molecules, which otherwise are contaminants, are washed out of the system using an appropriate buffer. The buffer is chosen so that ligand-ligate complex is not disrupted. Finally, the target molecule is eluted from the adsorbent with an appropriate medium (e.g., by varying the pH or ionic composition of the adsorption buffer) that weakens the bond between the complex pair. Competitive elution in which a counterligand in the eluting medium is used to displace the adsorbed ligate from the adsorbent surface is also employed.

Conventional immobilized ligand affinity separation processes suffer from several drawbacks that limit their range of application and also lead to reduced ligand utilization and product contamination. To begin with, the ligand must have at least two functionalities, one to enable attachment to the matrix and the other to enable interaction with the target molecule. Moreover, the immobilization procedure is tedious and can be time consuming. An equally serious drawback is the exposure of ligands to media that contain species other than the molecule of interest. These species adsorb (physically or specifically) and block active sites (fouling) of the ligand to target molecules. Drag and shear forces occurring in the medium often cause ligand leakage from the system and reduce system capacity. Mass transfer resistances also lead to long equilibration times. These problems make the application of conventional immobilized ligand affinity separation unsuitable for product isolation from media containing considerable levels of foreign matter as occurs in whole broth processing [5,6].

Several new affinity techniques have evolved over the last decade. Emphasis on improving the conventional process has centered on new immobilization strategies with a view to reducing mass transfer barriers, nonspecific adsorption, and, consequently, increased process selectivity and ligand utilization. Recent advances include affinity precipitation [7] in which the change in solubility of a soluble ligand upon attachment to a target molecule is utilized to effect the isolation of the latter, and affinity partitioning [8] in which a change in the partition coefficient of a ligand between a two-phase aqueous polymer solution upon attachment to a solute is harnessed to effect the separation of the solute. Other methods include membrane affinity separation [1], affinity based micellar extraction [9], continuous affinity recycle and extraction [11], and affinity electrophoresis [12]. These newer strategies too posses drawbacks. In affinity precipitation and affinity partitioning, one must process large volumes of solvent that may not always be friendly. In membrane affinity separation, fouling of ligand sites leads to reduced adsorption capacities of the system. More importantly, this strategy does not offer any improvement in ligand-coupling strategy. In micellar affinity processes the introduction of an additional step, namely, having to disrupt the micellar before retrieving the ligand-adsorbate complex, could lengthen processing time.

Recently, Daugulis et al. [13] introduced affinity microcapsules as an alternative to conventional immobilized ligand affinity separation systems (Figure 1). They successfully isolated bovine serum albumin from plasma medium using a blue dextran affinity ligand entrapped in poly-L-lysine-alginate microcapsules. However, they were unsuccessful when they attempted to use chitosan-alginate microcapsules for the isolation since the capsules disintegrated in the plasma medium. They concluded that chitosan-alginate microcapsules could not be utilized in the plasma medium to isolate albumin. Furthermore, they did not investigate the effect of capsule parameters on adsorption behavior. Adsorption constants useful for predictive purposes were not explored.

The microcapsule, which normally has a diameter of 50–600 μm, has a core that contains the affinity ligand in a fluid medium. This core fluid is enclosed within a semipermeable polymeric gel membrane. At the adsorption stage the affinity microcapsules are mixed with the medium containing the target molecule (Figure 1). During this process molecules small enough to traverse the semipermeable membrane diffuse into the core of the capsule. Larger molecules are prevented from going across the membrane. Among the molecules that diffused into

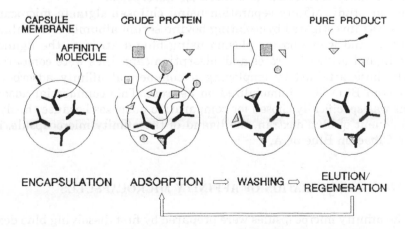

Figure 1. The concept of bioseparation using affinity microcapsules.

the capsule core, those that have affinity for the ligand, are adsorbed. At the washing stage the nonbound molecules are removed. Finally, the target molecules are retrieved at the elution stage by employing an appropriate medium to disrupt the ligand-adsorbate complex. The affinity microcapsule should be differentiated from gel-enclosed adsorbents described by Nigam et al. [5] where solid adsorbent particles were entrapped in gel matrices, or the system described by Kopperschlager et al. [14] where the ligand, blue dextran, was immobilized by fixation in cross-linked polyacrylamide gel.

Immobilizing affinity ligands in microcapsules holds several advantages over conventional techniques. The semipermeable membrane serves as a molecular sieve allowing only molecules of certain sizes, defined by the pore size of the outer semipermeable membrane, to traverse the membrane and interact with the ligand. Also, the differences in diffusion rates of diffusing molecules can be harnessed to improve product purity. These, in addition to the selectivity of the ligand, can result in a cascade of separation steps and improved overall process selectivity [15]. The pore size of the gel membrane can be altered by altering the capsule environment. This property of the membrane can be exploited to restrict transport through the membrane to molecules of certain sizes only during adsorption or speed-up desorption by opening up membrane pores during the elution cycle. For general ligands, such as dyes, this could be crucial for improving the selectivity. Further advantages of encapsulating affinity ligands include the protection the membrane offers the ligand from fouling, and shear and drag forces. Finally, the potential for higher adsorption capacities exists since the capsule volume rather than a bead surface, as is the case in conventional affinity separation, is utilized for adsorption.

In our study, affinity separation using chitosan-alginate microcapsules was investigated by isolating bovine serum albumin from saline solution and whole bovine plasma using blue dextran as the ligand. The ligand utilization rates and adsorption capacities were compared with those attained by employing a commercial affinity adsorbent Cibacron Blue 3GA immobilized on agarose. This commercial adsorbent was specifically chosen for comparative purposes because the dye component of blue dextran, the ligand of the affinity microcapsule, is also Cibacron Blue 3GA.

PREPARATION OF AFFINITY MICROCAPSULES

The affinity microcapsules were prepared by first dissolving blue dextran (0.2–2 g) and sodium alginate (0.6–2 g) in 100 ml water. The quantities of blue dextran and sodium alginate used depended on the ligand loading and the viscosity of the capsule core that was required for a particular experiment, respectively. The sodium alginate-blue dextran mixture (1–2 ml) was then extruded, using an electrostatic droplet generator, through a 22-gauge needle with a syringe pump into an aqueous solution of calcium chloride (3% w/v). The droplet generator consisted essentially of two electrodes, a positive and a ground electrode, from a high-voltage source attached to the syringe needle containing the blue dextran-alginate mixture and the collecting vessel, respectively. The droplets emerging from the needle were collected in a vessel that contained the calcium chloride solution. The size of the droplets and the rate of formation of droplets were varied by varying the operating voltage between 6 and 12 kV and the distance between the tip of the needle and the collecting vessel (3.5–4.5 cm). A detailed description of the electrostatic droplet generator was given by Bugarski et al. [17]. One minute after extrusion was completed, the formed calcium-alginate beads were washed in 0.5% w/v NaCl(aq) and subsequently transferred into an aqueous solution of chitosan (0.5% w/v) for 10 minutes to form the semipermeable membrane of the capsule. The capsules were finally thoroughly washed in 0.5% (w/v) sodium chloride and thereafter used for the adsorption studies.

A 0.5% (w/v) chitosan solution was prepared by measuring 0.5 g chitosan into 100 ml water in a beaker and adding 0.5 ml glacial acetic acid. Sodium nitrite, 0.1 g/1 g chitosan, was then added. The mixture was stirred till the chitosan was completely dissolved (pH of the resulting aqueous chitosan solution was 3.90). The chitosan dissolved faster when digested in this manner compared with the conventional approach described by Peniston and Johnson [16] where the chitosan is first dissolved in dilute acetic acid solution (aq) before the sodium nitrite is added to reduce the molecular weight of the chitosan. Chitosan solutions prepared in the manner described had a more consistent molecular weight distribution.

Blue dextran, dextran molecular weight 2×10^6, and bovine serum albumin (BSA) were used as ligand and solute, respectively. All were purchased from Sigma Chemical Co., St. Louis, MO. Sodium alginate, chitosan (practical grade, from crab shells), and sodium chloride used in the preparation of the capsules were also purchased from Sigma Chemical Co. Calcium chloride (anhydrous), glacial acetic acid, and sodium thiocyanate, all reagent grade, were obtained from Fisher Scientific, Fair Lawn, NJ. Sodium nitrite, reagent grade, was purchased from Aldrich Chemical Co. Inc., Milwaukee, WI. The dye reagent concentrate for assaying the protein was purchased from Bio-Rad Laboratories Ltd., Mississaugua, ON, Canada. Cibacron Blue 3GA (insolubilized on 4% beaded agarose) and lyophilized bovine plasma were purchased from Sigma.

The affinity adsorbent, Cibacron Blue 3GA immobilized on agarose, as purchased from the manufacturer, contained lactose as stabilizer, which must be removed prior to use. This was done by washing 0.2 g of dry adsorbent in 50 ml distilled water and filtering through a filter paper. The gel was then collected from the filter paper and used in the protein adsorption studies (the gel swelled to about 1 ml upon immersion in water).

ADSORPTION STUDIES

The affinity microcapsules, 2 ml (capsule diameter 300–530 μm), were incubated in 50 ml of saline solution (0.5% w/v NaCl, pH = 6.5 $\pm$ 0.5) containing dissolved bovine serum albumin (0.25–10 mg/ml) at room temperature in a stirred tank. The mixture was constantly stirred with an impeller. Samples were taken for analysis by first stopping the mixing and allowing the capsules to settle to the bottom of the stirred tank. The protein solution (100 μl) was then taken for analysis. The change in protein concentration was determined using the Bio-Rad Assay [18]. The equilibrium adsorption of the affinity capsules was identified as that stage in the adsorption process where there was no further change in concentration of protein solution in the stirred tank. The actual equilibrium protein adsorption was determined by a mass balance. The same procedure was used to study the adsorption of bovine serum albumin onto Cibacron Blue 3GA immobilized on agarose. Washed adsorbent, 0.2 g (1 ml gel), was incubated in 50 ml of protein solution, and the equilibrium position was tracked by occasionally taking samples from the mixture and measuring the protein concentration. This was continued until there was no longer any change in the albumin concentration.

To assess the potential of the affinity microcapsule concept, a universal measure was needed so that a quantitative comparison could be made against established techniques. Most researchers use the adsorption capacity to characterize the efficiency of a given affinity separation process. In particular, the adsorption capacity measured in mg adsorbed protein/ml adsorbent (or mole adsorbed protein/ml adsorbent) is very confusing when one wishes to compare different affinity adsorbents. A milliliter of agarose gel adsorbent compares poorly with a milliliter of silica bead adsorbent (or a milliliter of affinity microcapsules for that matter). Even though these might have the same ligand loading, they will invariably contain different amounts of solid material other than the ligand. Therefore, using the adsorption capacity as a comparative measure obscures the real effectiveness of the ligand and hence the comparative strengths of different systems. Ligand utilization measured in moles adsorbate/moles ligand present on the other hand eliminates this ambiguity. It measures the extent of actual interaction between the ligand and the adsorbate in a given system. In practice, the object of the affinity separation process is to provide conditions for effective interaction between the ligand and the target molecule. The extent of this interaction (ligand utilization) is therefore a better measure of the comparative strengths of different affinity separation systems [1] and is used in this work as a basis for comparing affinity microcapsules with affinity adsorption using Cibacron Blue 3GA immobilized on agarose.

Bovine plasma was obtained in lyophilized state and reconstituted with 10 ml deionized water. The resulting plasma was further diluted, 100%, with standard buffer pH = 5. The chitosan-alginate capsules are stable at this pH. Affinity microcapsules (average diameter of 530 μm, total volume 2 ml), in the case of the microcapsule studies and 0.2 g of washed Cibacron Blue 3GA immobilized on agarose in the case of studies with the commercial adsorbent, were placed in 40 ml of the reconstituted plasma in a stirred tank. The mixture was stirred for 6 h with an impeller. The affinity adsorbents were then filtered and washed with water and then finally washed with the adsorption buffer on a filter paper. Subsequently, the adsorbents were placed in the stirred tank and the adsorbed albumin eluted with 40 ml 0.2 M sodium thiocyanate. Aliquots, 2 ml, were taken at each stage of the process for SDS-PAGE analysis.

During capsule preparation, loss of ligand increased with decreasing capsule diameter (Figure 2). For example, at capsule diameter of 300 μm, 49% of the ligand was lost, whereas at capsule diameter of 530 μm only 23% of the ligand was lost. The loss was due mainly to the dissolution of the ligand into the aqueous medium during the formation of calcium alginate-blue dextran bead from sodium alginate-blue dextran droplets. Since the process is purely diffusional, the smaller capsules with an overall larger surface area suffer a greater ligand loss than larger capsules with a smaller overall surface area. In the case of the Cibacron Blue 3GA immobilized on agarose the ligand loading of the adsorbent was taken as that reported by the manufacturer. The percentage ligand loss, LS, was computed as

$$LS = 100 \times (L1 - L2)/L1 \tag{1a}$$

where $L1$ is the amount of ligand, in milligrams, in the original mixture used in the preparation of the affinity capsules, and $L2$ is the actual amount of ligand that was retained by the capsules prepared from the mixture. The error associated with the calculated percentage ligand loss was $\pm 4.2\%$.

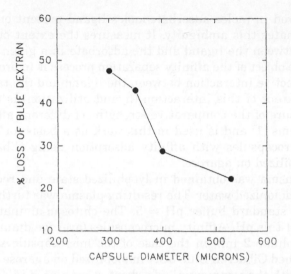

Figure 2. Ligand loss as a function capsule size during synthesis of affinity microcapsules (ligand loading = 20 mg/ml; capsule average diameter = 530 μm; viscosity of capsule core fluid = 0.006 Pa·s; volume of capsules = 2 ml; volume of protein solution = 50 ml).

In the absence of blue dextran there was no significant decrease in bulk protein concentration (Figure 3). The protein concentration decreased from an initial value of 0.25 mg/ml to a value of 0.248 mg/ml at equilibrium (a change of only 0.012 mg/ml). However, there was a dramatic decrease in the bulk protein concentration as the blue dextran loading increased. The equilibrium bulk protein concentration (i.e., at 8 h) was 0.15 mg/ml at a ligand loading of 2 mg/ml capsule and decreased to 0.02 mg/ml at a ligand loading of 20 mg/ml capsule. The concentration of adsorbed protein was 5.75 mg/ml capsule at the higher loading. The foregoing results confirm that the change in bulk protein concentration was due to protein-ligand interaction rather than to physical attachment of protein to the outer layer of the capsules. Holland [19] reported the successful use of chitosan gel beads in chromatographic separations for the isolation of bovine serum albumin. In spite of the fact that the outer layer of the affinity microcapsule was made of chitosan, as the foregoing results suggest this outer chitosan layer has no significant influence on the adsorption process. It must, however, be pointed out that the quantity of chitosan on the microcapsules is minimal and compares poorly with a situation where an entire column is packed with chitosan gel as was the case in the study reported by Holland.

Adsorption Isotherms

The equilibrium isotherms for the adsorption of bovine serum albumin by microcapsules containing Cibacron Blue 3GA dextran and Cibacron Blue 3GA immobilized on agarose are shown on Figure 4. The hyperbolic nature of the curves suggests that these are of the Langmuirean type [20]. The interaction between the albumin and the ligand for either system can therefore be approximated as

$$L + P \overset{k_1}{\underset{k_2}{\rightleftharpoons}} LP \tag{1}$$

where L is the ligand, P is the protein, and LP is the ligand-protein complex.

Based on Equation (1) the rate of protein adsorption onto the adsorbents can be described as [21]

$$\frac{\partial c_{lp}}{\partial t} = k_1 c_p (c_{lm} - c_{lp}) - k_2 c_{lp} \tag{2}$$

where c_p is the protein concentration in the stirred tank at any time t, c_{lm} is the maximum ligand site concentration (adsorption capacity of

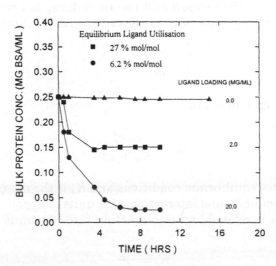

Figure 3. Effect of ligand loading on bulk protein adsorption (capsule diameter = 530 μm; capsule core viscosity = 0.006 Pa·s; volume of capsule = 2 ml; volume of protein solution = 50 ml).

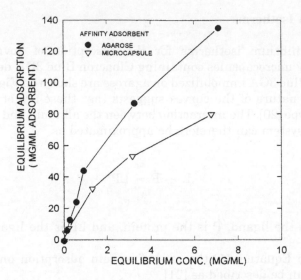

Figure 4. Isotherms for the adsorption of albumin onto affinity microcapsules containing Cibacron Blue 3GA (capsule average diameter = 530 μm; ligand loading = 1.4 μmol/ml; volume of capsules = 2 ml; capsule core viscosity = 0.006 Pa·s) and agarose containing Cibacron Blue 3GA on 4% agarose (gel bead diameter) = 10–100 μm; ligand loading = 2.8 μmol/ml; gel volume = 1 ml; volume of protein solution = 50 ml).

the adsorbent), c_{lp} is the concentration of the ligand-protein complex, and k_1 and k_2 are the forward and reverse interaction rate constants with units of ml mg^{-1} and min^{-1}, respectively. At equilibrium,

$$\frac{\partial c_{lp}}{\partial t} = 0 \tag{3}$$

and Equation (2) reduces to

$$c_{lp}^* = \frac{c_{lm} c_p^*}{K_d + c_p^*} \tag{4}$$

where * denotes equilibrium conditions and K_d is the dissociation constant of the protein-ligand interaction with units of mg/ml. The dissociation constant is related to the interaction rate constants by

$$K_d = \frac{k_2}{k_1} \tag{5}$$

Equation (4) was linearized and fitted to the equilibrium data for both systems. The unknown parameter's adsorption capacity of the adsorbant, c_{lm}, and dissociation constant, K_d, were subsequently determined from the plots (Table 1). Based on adsorption capacities measured in mg/ml, it would initially appear that the commercial adsorbent, Cibacron Blue 3GA immobilized on agarose, is better than the affinity microcapsule (189 mg protein/ml adsorbent for the commercial adsorbent compared with 103 mg protein/ml adsorbent for the affinity microcapsule). However, based on real interaction between ligand and protein molecules, measured as ligand utilization, the affinity microcapsules show comparable adsorption compared with the commercial adsorbent (108% μmol protein/μmol ligand to 90% μmol protein/μmol ligand for the commercial adsorbent) (Table 1).

Effect of Capsule Parameters on Equilibrium Behavior of the Affinity Microcapsule

The effect of capsule diameter on bulk protein concentration was investigated (Figure 5). Protein adsorption attained a minimum prior to equilibrium. A similar phenomena was observed by Nigam et al. [5] for the adsorption of proteins and enzymes onto gel-entrapped adsorbents from a two-component mixture. They explained the phenomena as due to the displacement of already adsorbed smaller enzyme molecules by slower diffusing larger BSA molecules. The system in our work, how-

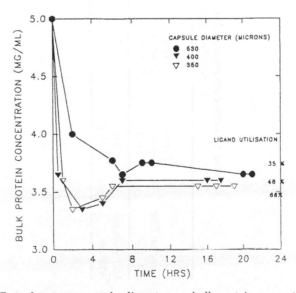

Figure 5. Effect of average capsule diameter on bulk protein concentration (ligand loading = 1.4 μmol/capsule; volume of capsules = 2 ml; capsule core viscosity = 0.006 Pa·s; volume of protein solution = 50 ml).

Table 1. Adsorbent properties and adsorption constant of Cibacron Blue 3GA-dextran in microcapsules and Cibacron Blue 3GA on agarose.

Cibacron Blue 3GA	Particle Diameter (μm)	Ligand Loading (mol/ml)	Volume of Adsorbent (ml)	Adsorption Capacity c_{im} (mg/ml)	Maximum Ligand Utilization (% ± S.D.)	Dissociation Constant, K_d (mg/ml)
Microcapsules	530	1.4	2	103 ± 13.1	108 ± 7.8	2.17
Agarose	10–100	2.8	1	189 ± 22.2	90 ± 15.8	3.40

ever, contained a single protein, and yet the minimum was still evident. It is more likely that the adsorption behavior in our studies was due to the changes in structure of the proteins upon adsorption as explained by Norde [22]. First, the protein attaches to the ligand at an active site, and then gradually it adjusts its structure to attach other parts of its body to the adsorbent. As this happens, nearby proteins are displaced off the adsorbent. Hence the increase in the bulk protein concentration (i.e., decrease in the adsorbed protein concentration). This phenomena is not observable at low initial protein concentrations (i.e., < 1.0 mg/ml, see Figures 3 and 6) where the incidence of active ligand site saturation with protein is minimal, hence protein displacement as a result of stretching of nearby molecules is greatly reduced or nonexistent.

An increase in equilibrium adsorption with decreasing capsule diameter was observed (Figure 5). For example, at capsule diameter of 530 μm the ligand utilization at equilibrium was 35% with this value climbing to 66% at capsule diameter of 350 μm. This result suggests that the capsule surface area plays a crucial role in the ligand utilization of the affinity microcapsule. It should be noted that this behavior is not attributable to a decrease in mass transport resistance with increasing capsule surface area (decreasing capsule diameter) because this would only affect the rate of attainment of equilibrium and not the

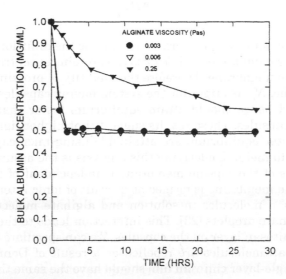

Figure 6. Effect of capsule core viscosity on bulk protein concentration at the adsorption stage (average capsule diameter = 530 μm; ligand loading = 1.4 μm/ml capsules; capsule volume = 2 ml; volume of protein solution = 50 ml).

equilibrium adsorption. This trend in adsorption is most probably due to the utilization of both capsule core volume as well as the membrane core fluid interface (i.e., gel layer) for adsorption. Protein molecules diffuse into the capsule volume to adsorb onto the ligands floating in the core fluid while at the same time diffusing protein molecules adsorb onto ligands trapped at the gel interface formed during membrane formation. With time, it is expected that the interface would become saturated before the capsule core is saturated. We can speculate that access to the capsule core may then be sterically hindered by proteins attached to the interface. Consequently, the final capsular adsorption will be determined by capsule core protein saturation and the protein saturation at the interface. The membrane core fluid interface would saturate first with proteins. When this happens, access to the ligand in the core fluid might be blocked. It is, therefore, expected that a system having smaller capsules (larger overall interfacial area) should have higher equilibrium adsorption than larger capsules (smaller overall interfacial area), hence, the observed trend. The change in rate of attainment of equilibrium with capsule size follows the same trend. The increase in overall surface area of the capsule with decrease in size directly influences the flux of protein into the capsule. The protein mass transfer rate (N_p) is defined according to Fick's first law as

$$N_p = AD_m \frac{dc}{dr} \tag{6}$$

where A is the total capsule surface area, D_m is the diffusion coefficient in the capsule membrane, and dc/dr is the protein concentration gradient across the membrane. Therefore the quantity of protein traversing the membrane, N_p, is larger for the system having capsules of smaller diameter and hence should attain equilibrium faster than a system with larger capsules. These results suggest that higher ligand utilizations and faster equilibrium are attainable using microcapsules having smaller diameter. Underlying this analysis is the assumption that the thickness of the capsule membrane is independent of the capsule diameter. The membrane is formed as a result of ionic interactions between chitosan molecules in solution and alginate molecules in the calcium alginate droplets [23]. This interaction leads to the formation of a single chitosan layer on the capsules. We can speculate that subsequent chitosan molecules are repelled as a result of Donnan effects. Thus this single-layer chitosan film should have the same thickness independent of the capsule diameter.

When the viscosity of the sodium-alginate solution employed in making the capsules was doubled from 0.003 to 0.006 Pa·s, there was no apparent change in either the rate of attainment of equilibrium or equilibrium adsorption (Figure 6). However, when the viscosity was increased by two orders of magnitude (from 0.006 to 0.25 Pa·s), the differences in both the rate of attainment of equilibrium and equilibrium adsorption became marked. This behavior was probably due to the fact that at the low viscosity end the process was limited by transport through the membrane. Consequently, the change in viscosity did not have much effect on either the rate of attainment of equilibrium or on the equilibrium adsorption. As the core viscosity increased, however, transport resistance in the core fluid became significant, becoming perhaps, more important than the membrane resistance, hence, the change in rate of attainment of equilibrium. For example, at an alginate viscosity of 0.25 Pa·s it took 24 h to reach equilibrium, whereas at a core viscosity of 0.006 Pa·s it took only 2 h. It must be noted that the diffusion coefficient D_{AB} is related to the viscosity, μ, and the molecular weight of the diffusing molecule, M_A, [24],

$$D_{AB} = 9.4 \times 10^{-15} \frac{T}{\mu(M_A)^{1/3}} \tag{7}$$

where T is the absolute temperature of the medium. According to Equation (7), as the capsule core viscosity increases, the diffusion coefficient and, hence, the rate of mass transport in the capsule core decreases. Thus, the time to attain equilibrium is expected to increase at higher viscosities as the experimental results demonstrate (Figure 6). Also, as the alginate mass in the capsule increases, physical attachment of alginate molecules to the ligand may increase site blockage of the ligand, and hence, there may be an accompanying decrease in ligand utilization. Based on the foregoing results, higher ligand utilizations and faster equilibrium can thus be achieved by using affinity microcapsules with a low capsule core fluid viscosity. It should, however, be noted that decreasing the capsule core viscosity (low capsule core alginate concentration) may result in increased loss of ligand from the capsules because longer times might be needed to form the capsules). In fact, at the lowest sodium-alginate concentration tested, 0.6% w/v and a blue dextran concentration of 2% w/v, the microcapsules did not form at all. Consequently, the lowest capsule core viscosity (sodium-alginate concentration) that could be employed would be determined by the ability to form microcapsules from the adsorbent mixture.

A direct comparison between the adsorption behavior of the microcapsules containing blue dextran and the commercial adsorbent, Cibacron Blue 3GA immobilized on agarose, is shown in Figure 7. The affinity microcapsule exhibited higher adsorption for albumin than the

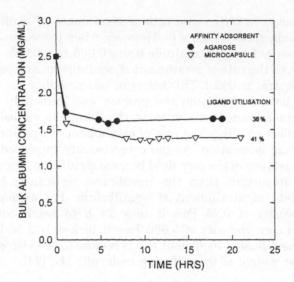

Figure 7. Comparison of different ligand immobilization schemes at the adsorption stage (average capsule diameter = 530 μm; capsule ligand loading = 1.4 μmol/ml capsule; volume of capsules = 2 ml; diameter of commercial adsorbent = 10–100 μm; ligand loading = 2.8 μmol/ml adsorbent; volume of adsorbent = 1 ml; volume of protein solution = 50 ml).

commercial adsorbent as shown by the lower residual bulk protein concentration. It is worth noting that the slope of the adsorption curve for the affinity microcapsules in the first 2 h of the adsorption process is comparable to that of the commercial adsorbent, indicating that the adsorption rates are similar. Moreover, the change in bulk protein concentration, for either adsorbent, beyond the first 2 h of adsorption is very small. Since it takes more than 5 h to attain equilibrium for either product, and in the light of the small changes in adsorption beyond the first 2 h of the adsorption process, in an industrial setting the adsorption process could thus probably be terminated after only 2 h.

The variation in ligand utilization with initial bulk protein concentration is shown in Table 2. In all cases the microcapsule, corrected for ligand loss, showed consistently higher ligand utility. The higher adsorption of the affinity microcapsule compared with the conventional adsorbent is probably due to the improved mobility attendant to immobilizing the ligand in a fluid medium. The improved mobility results in reduced steric hindrance of ligand sites and, hence, increased utility of the ligand active sites. The errors associated with the measurement of the bulk protein concentrations and the calculated adsorbed amounts of protein were estimated as ±0.049 and ±2.48, respectively, at a 95% confidence interval.

Unlike conventional immobilized ligand affinity separation, the affinity microcapsule offers several easy ways of varying process parameters. The capsule parameters, viz., capsule ligand loading, capsule diameter, viscosity of capsule core fluid, membrane thickness, and membrane pore structure, can all be designed to effect changes in the separation strategy and also to get an insight into the kinetic behavior of the adsorption and desorption processes. The ability to vary these features of the microcapsule makes it an attractive alternative to the conventional affinity adsorbents. Although the first three will impinge directly on equilibrium adsorption and the rate of attainment of equilibrium, the last two are expected to affect only the rate of attainment of equilibrium.

The chitosan-alginate semipermeable membrane is formed as a result of electrostatic interactions between opposite ions on the alginate and chitosan macromolecules [25]. Once the membrane layer is formed, subsequent attachment of chitosan molecules may be presented as a result of electrostatic repulsion between like charges on the reacted chitosan molecules and those in solution. Consequently, the thickness of the semipermeable chitosan-alginate membrane is expected to be independent of the membrane formation reaction time. Indeed, the results of Polk et al. [26] showed that the loss of blue dextran and albumin from the capsules made with chitosan and alginate was independent of membrane formation reaction times greater than 10 minutes. King et al. [27] studied the effect of reaction time on permeability for the capsules formed from poly-L-lysine and sodium alginate with membrane thicknesses varying from 5 to 10 μm. They found that for this system the permeability is affected by the reaction time. Thus, for such systems the real possibility exists to use the degree

Table 2. *Effect of initial bulk protein concentration on ligand utilization of affinity adsorbents (ligand loading = 2.8 mol per 2 ml capsules or 1 ml agarose; capsule diameter = 530 μm; capsule core viscosity = 0.006 Pa·s; volume of protein solution = 50.0 ml).*

Initial Protein Concentration (mg/ml)	Equilibrium Ligand Utilization % (mole bound protein/mole Cibacron Blue 3GA) ($\pm$S.D.)		
	Affinity Microcapsule	Affinity Microcapsule[a]	Cibacron Blue 3GA on Agarose
0.25	5 $\pm$ 0.8	6 $\pm$ 1.1	4 $\pm$ 0.8
0.50	6 $\pm$ 0.8	8 $\pm$ 1.1	7 $\pm$ 0.8
2.50	26 $\pm$ 1.9	33 $\pm$ 2.6	23 $\pm$ 15.8
5.00	37 $\pm$ 7.8	47 $\pm$ 10.3	36 $\pm$ 15.8
10.00	74 $\pm$ 7.8	95 $\pm$ 10.3	73 $\pm$ 7.8

[a]Corrected for ligand loss during capsule preparation.

of cross-linking to change both membrane thickness and pore structure and, hence, the membrane permeability. On the other hand, the permeability behavior of chitosan-alginate membranes also depend on the stimuli in the capsule's external environment such as pH and ionic concentration and ionic composition of surrounding media [28,29]. It is worth noting that these same parameters, namely, pH, ionic concentration, and composition of the medium, also affect protein adsorption.

Isolation of Albumin from Bovine Plasma

The SDS-PAGE results showed that the microcapsule containing blue dextran could be successfully used to isolate albumin from diluted bovine plasma (Figure 8). The bands in Lanes C1 and C2 are the eluted albumin (MW 66 kD) from the affinity microcapsules and Cibacron Blue 3GA immobilized on agarose, respectively. Both adsorbents were effective in isolating albumin. The intensity of the adsorption bands suggest that the microcapsule has higher albumin adsorption capabilities than the commercial adsorbent. Lanes B1 and B2 are runs of samples taken after extensive washing of the capsules after the adsorption stage. Clearly, there is no albumin in these lanes. Consequently, the eluted protein must have come from the adsorbed albumin.

In a previous study from our laboratory, Daugulis et al. [13] attempted to isolate albumin from whole serum plasma using the chitosan-alginate microcapsules as well as poly-L-lysine-alginate capsules with blue dextran as the ligand. Although the poly-L-lysine-alginate capsules were stable, the chitosan-alginate capsules were not. The capsules dissolved in plasma. Consequently, it was not possible, at that time, to carry out the albumin recovery process with the chitosan-alginate microcapsules. The instability of the capsules could be attributed to the ionic composition as well as the pH of the plasma medium. These need to be manipulated to a composition of the plasma medium where the capsules are stable. That is the primary reason why we diluted whole serum plasma (100% dilution) with standard buffer (pH = 5). Chitosan-alginate capsules were stable in this buffered plasma, and, consequently, albumn isolation could be achieved.

CONCLUSIONS

Chitosan-alginate affinity microcapsules have been successfully employed for the recovery of bovine serum albumin from saline medium and whole bovine plasma using blue dextran as the affinity ligand. Experimental adsorption data for the microcapsule and a commercial adsorbent, Cibacron Blue 3GA immobilized on agarose, were

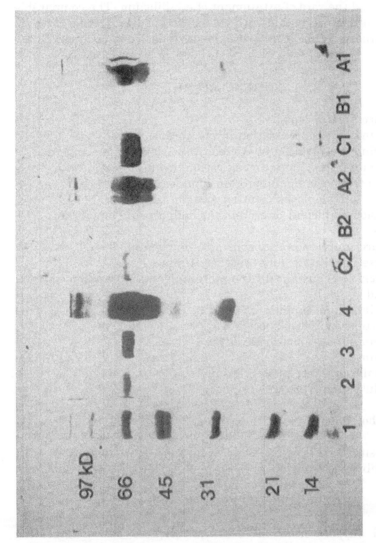

Figure 8. SDS-PAGE run of samples from different stages of albumin isolation from diluted bovine plasma: Lanes A1–C1 and A2–C2 represent runs of samples from the adsorption, washing and elution stages of the blue dextran microcapsules and Cibacron Blue 3GA immobilized on agarose, respectively. Lane 1 is the molecular weight marker; Lane 2 is a sample from precipitated protein during dilution of whole plasma. Lane 3 is commercial albumin. Lane 4 is diluted plasma.

comparable and showed that the affinity microcapsule concept is potentially a viable industrial separation tool. The capsule diameter, capsule core viscosity, and capsule ligand loading all influenced equilibrium adsorption and the rate of attainment of equilibrium. The comparable ligand utilization attainable with the microcapsule, together with its ease of synthesis, make it an attractive alternative to commercial adsorbents.

NOMENCLATURE

A surface area of capsule
c protein point concentration in membrane, mg/ml
c_p protein concentration in stirred tank at any time t, mg/ml
c_{lm} maximum ligand site concentration, mg/ml
c_{lp} concentration of ligand-protein complex, mg/ml
D_{AB} binary diffusion coefficient for system A-B, cm²/s
D_m diffusion coefficient of protein through capsule membrane, cm²/s
k_1 forward interaction rate constant, ml/mg·min
k_2 reverse interaction rate constant, 1/min
K_d dissociation constant for the protein-ligand interaction, mg/ml
L   ligand
LP   protein-ligand complex
M_A molecular weight of species A
N_p protein mass transfer rate, mg/s
P   protein
r spacial coordinate in the capsule
T absolute temperature

Greek symbols

μ viscosity
*   equilibrium conditions

ACKNOWLEDGEMENTS

Financial support from the Natural Sciences and Engineering Research Council of Canada is gratefully acknowledged.

REFERENCES

1. Brandt, S., Goffe, R. A., Kessler, S. B., O'Connor, J. L. and Zale, S. E. 1988. "Membrane-Based Affinity Technology for Commercial Scale Purifications," *Biotechnology*, 6:779.

2. Parikh, I. and Cuatrecasas, P. 1984. "Affinity Chromatography," *Chem. and Eng. News*, (Aug. 26th):17.

3. Porath, J. and Belew, M. 1983. In *Affinity Chromatography and Biological Recognition* (eds. Chaiken, I. M. et al.), Academic Press, San Diego, p. 173 .

4. Jansen, J. C. 1984. "Large-Scale Affinity Publication–State of the Art and Future Prospects," *Trends in Biotechnology*, 2:31.

5. Nigam, S. C., Sakoda, A. and Wang, H. Y. 1988. "Bioproduct Recovery from Unclarified Broths and Homogenates Using Immobilised Adsorbents," *Biotechnol. Prog.*, 4:166.

6. Frez, A. K., Gustafsson, J. and Hedman, P. 1986. "Recovery of β-Galactosidase by Adsorption from Unclarified *Escherichia coli* Homogenate," *Biotech. Bioeng.*, 28:133.

7. Homma, T., Fuji, M., Mori, J., Kawakami, T., Kuroda, K. and Taguchi, M. 1993. "Production of Cellobiose by Enzymatic Hydrolysis. Removal of Beta-Glucosidase from Cellulase by Affinity Precipitation Using Chitosan," *Biotech. Bioeng.*, 41:405.

8. Johansson, G. 1987. "Dye-Ligand Aqueous Two-Phase Systems" in *Reactive Dyes in Protein and Enzyme Technology* (eds. Clonis, Y. D., Atkinson, A., Bruton, C. J. and Lowe, C. R., Stockton Press, NY, p. 101.

9. Paradkar, V. M. and Dordick, J. S. 1993. "Affinity-Based Reverse Micellar Extraction and Separation (ARMES): A Facile Technique for the Purification of Peroxidase from Soybean Hulls," *Biotechnol. Prog.*, 9:199 .

10. Woll, J. M., Hatton, T. A. and Yarmush, M. L. 1989. "Bioaffinity Separations Using Reverse Micellar Extraction," *Biotechnol. Prog.*, 5:57 .

11. Afeyan, N. B., Gordon, N. F. and Cooney, L. C. 1989. "Continuous Affinity Recycle Extraction: A Novel Protein Separation Approach," *J. of Chromatography*, 478:1.

12. Shimura, K. 1990. "Progress in Affinophoresis," *J. of Chromatography*, 510:251 .

13. Daugulis, A. J., Goosen, M. F. A., Wight, C. P. and Chesney, K. E. 1990. "Affinity Separation Using Microcapsules," *SCI International Conference on Separation for Biotechnology*, Univ. of Reading, U.K. (EL Series Publ. Ltd.), Sept.

14. Kopperschlager, G., Bohme, H-J. and Hofman, E. 1982. "Cibacron Blue F3G-A and Related Dyes as Ligands in Affinity Chromatography," *Advances in Biochemical Engineering/Biotechnology*, 25:103.

15. Nigam, C. S., Siapush, A. R. and Wang, H. Y. 1990. "Analysis of Bioproduct Separation Using Gel-Enclosed Adsorbents," *AIChE Journal*, 36:1239 .

16. Peniston, Q. P. and Johnson, E. L. 1975. "Process for Depolymerization of Chitosan," U.S. Patent No. 3,922,260.

17. Bugarski, B., Poncelet, D., Neufeld, R. J., Li, Q., Vunjak, G. and Goosen, M. F. A. 1994. "Electrostatic Droplet Generation: Mechanism of Polymer Droplet Formation," *AIChE Journal*, 40(6):1026.

18. Bradford, M. 1976. "A Rapid and Sensitive Method for the Quantitation of

Microgram Quantities of Protein Utilizing the Principle of Protein-Dye Binding," *Anal. Biochem.,* 72:248.

19. Holland, C. R. 1988. In *Chitin and Chitosan* (eds. Gudmund, S. K., Thorleif, A. and Sandford, P.), Elsevier Applied Science, p. 558.

20. Liapis, A. E., Anspach, B., Findley, M. E., Davies, M. Hearn, M. T. W. and Unger, K. K. 1989. "Biospecific Adsorption of Lysozyme onto Monoclonal Antibody Ligand Immobilised on Porous Silica Particles," *Biotech. and Bioeng.,* 34:466.

21. Chase, H. A. 1984. "Prediction of the Performance of Preparative Affinity Chromatography," *J. of Chromatography,* 297:179.

22. Norde, W. 1986. "Adsorption of Proteins from Solution at Solid-Liquid Interface," *Advances in Colloid and Interface Science,* 25:267.

23. McKnight, C. A., Ku, A. and Goosen, M. F. 1988. "Synthesis of Chitosan-Alginate Microcapsules Membranes," *J. of Bioactive and Compatible Polymers,* 3:334.

24. Polson, A. 1950. "Some Aspects of Diffusion in Solution and a Definition of a Colloidal Particle," *Journal of Physical and Colloidal Chemistry,* 54:649.

25. Dunn, E. J., Zhang, X., Sun, D. and Goosen, M. F. A. 1988. "Synthesis of *N*-(Aminoalkyl) Chitosan for Microcapsules," *J. Applied Polymer Science,* 50:353.

26. Polk, A., Amsden, B., Yao, K. D., Peng, T. and Goosen, M. F. A. 1994. "Controlled Release of Albumin from Chitosan-Alginate Microcapsules," *J. Pharmaceutical Sciences,* 83(2):178.

27. King, G. A., Daugulis, A. J., Faulkner, P. and Goosen, M. F. A. 1987. "Alginate-poly-L-lysine Microcapsues of Controlled Membrane Molecular Weight Cut-off for Mammalian Cell Culture Engineering," *Biotechnology Progress,* 3:4.

28. Shah, B. G. and Barnett, S. M. 1992. "Hyaluronic Acid Gels," in *Polyelectrolyte Gels* (eds. Harland, R. S. and Prud'homme, R. K.), ACS Symposium Series 480, p. 116.

29. Grimshaw, P. E., Grodzinsky, A. J., Yarmush, M. L. and Yarmush, D. M. 1989. "Dynamic Membranes for Protein Transport: Chemical and Electrical Control," *Chem. Eng. Sci.,* 44:827.

CHAPTER 15

Structure and Isolation of Native Animal Chitins

MARIA L. BADE

Center for Cancer Research
Massachusetts Institute of Technology
Cambridge, MA 02139

INTRODUCTION

Chitin has been known to science for almost two centuries, but elucidation of its precise behavior via-à-vis specific enzymes has been slow. A splendid review of the literature of chitin chemistry and related matters up to about 1950 can be found in Reference [1], now unhappily out of print. In what follows, I shall cite from Richards's book, augment and reinterpret data where this seems appropriate, and show some of the extensive work done in my laboratory on the structure of chitins that are intended for enzyme substrates. When this work was first undertaken, a hoped-for bonus was that achieving construction of a "good" chitin substrate would lead to a better behaved constituent in the many practical applications proposed for chitin and chitosan, and this hope was realized (Bade, manuscript in preparation).

CHITIN: CHEMICAL IDENTITY

Chitin was discovered in 1811 by Braconnot who called it fungine [2]. The name chitin was conferred by Odier in 1823 [3] "for what has subsequently been found to be the same compound in insects" [1]. For many years after that, there was controversy about whether or not chitin was identical with cellulose, although as early as 1843 Payen had noted that chitin contains nitrogen. Ledderhose in his 1876 and later pub-

255

lications [4] of the results of acid hydrolysis of the remains of a lobster dinner with his uncle, von Helmholtz, in which a shedder was consumed, as well as hydrolysis of beetle elytra, reported obtaining a crystalline compound he named glucosamine; he also collected acetic acid but failed to draw the inference that acetyl might have been linked to the amine group *in situ*. With modern hindsight regarding the mechanism of acid hydrolysis, it is clear that following hydrolytic cleavage of the *N*-acyl linkage, the primary amine of the deacetylated chitin monomer still part of the polymer chain is hydrogenated to a full positive charge; this blocks hydronium ion attack on the immediately adjacent glycosidic link so that ring cleavage is encouraged instead. Ledderhose might therefore not have obtained stoichiometric amounts of the two hydrolytic products named. But as recently as 1921, Levene suggested that since the monomeric sugar obtained in acid hydrolysis from chitin might be mannosamine converted to glucosamine by Walden inversion during acid degradation, the sugar in chitin ought to be called chitosamine until the matter was more decisively settled [5]. Galactose, fucose, glucose, and mannose have all been reported as constituents (of chitosan) from Coelenterates, a Polyzoan, and other marine organisms [6], although the color reaction typical of chitosan was seen only subsequent to 24 h boiling in 5% KOH followed by a 50-minute reaction with saturated alkali at 160°C which, as noted below, will certainly deacetylate many *N*-acetylglucosamine monomers of any polymeric chitin that may have existed at the outset. As will be suggested below, there is reason to believe that even regular α-chitin might contain, within the primary sequence, another, highly reactive, nonnitrogenous constituent that may or may not be a sugar.

CHITIN: STRUCTURE

One unresolved controversy with respect to chitin structure is, then, whether there are other components involved in the primary β-1,4-linked sequence of nitrogen-containing sugars that form the polymer. Other questions are whether there are recoverable subunits intermediate between "chitin" particles as customarily isolated, and individual β-1,4-linked chains; whether chitins, of whatever biological origin, are identical; whether X-ray data giving α, β-, and occasionally γ-patterns are correctly interpreted as denoting parallel or antiparallel chain arrangement or a variant thereof [7]; and whether X-ray data may or may not distinguish between fibrous and particulate growth of chitin structures (this has engendered a discussion in the literature, but birefringence studies should show exactly what form of chitin is being synthesized).

Are there subunits within exoskeletal chitin and if so, can they be recovered? When care is taken not to disturb the native array of chitin molecules in isolation of partially purified chitinous animal structures, assembly of chitin chains into highly elongated and discrete micro- and macrofibrils is apparent from photomicrographs [Figure 1(a), 1(b)] [7a]. Microfibrils have previously been discussed [8, 8a].

Disturbances in the orderly array depicted in Figure 1(a) and (b) can arise in several ways during purification of chitin from animal structures, and most of the altered forms obtained thereby cannot thereafter be restored to native structure and/or spacing between microassemblies. In view of the pictures reproduced above and other data to be presented, it is proposed that in the case of chitin, several orders exist *in situ* that are analogous to those well known from the description of native proteins. Thus, the linear sequence of sugar monomers linked by covalent β-1,4-glycosidic bonds is termed the primary structure. Assembly into microfibrils, i.e., primary chitin chains linked in three dimensions through multiple hydrogen bonds, would be the secondary structure. Note that during biosynthesis, assembly into microfibrils is somehow limited (see Reference [8] for discussion). Rudall and Kenchington [7] estimated an array of twenty-one chains per microfibril, but they also suggested an antiparallel arrangement of individual chains within the macrofibril of the most common form of animal chitin, viz., α-chitin, and it is not altogether clear that the antiparallel construction is correct. (For one thing, it makes construction of a rational model for chitin synthesis very difficult.) Macrofibrils and sheets composed of them are proposed to be tertiary and quaternary structures, respectively: From Figure 1(b) it is apparent that protein surrounds the microfibrils and that these assemblies are structured into macrofibrils. Again, macrofibrils are limited in diameter by an unknown mechanism. Parallel arrays of macrofibrils are, in turn, assembled into chitin-protein sheets.

Bouligand [9] has interpreted the usual appearance of EM photographs of chitinous insect cuticle at fairly low magnification in terms of successive quaternary sheets being deposited in a skewed manner relative to the direction of the adjacent sheets directly above and below. The appearance of fibers cut in a glancing manner, next to fibers that are laid down in parallel, is consistent with that interpretation [cf. Figure 1(b)]. Others have spoken of "helicoidal layers" [8]. Figure 2 shows an EM picture of crab endocuticle in sagittal section, and by Bouligand's theory, successive sheets are laid down so that in seemingly parallel sheets, fibers making up the sheet are running at a 180° angle to the 170° and 190° sheets, i.e., if the first sheet is at 0°, successive sheets are seen to turn and be cut at a glancing angle, each approx-

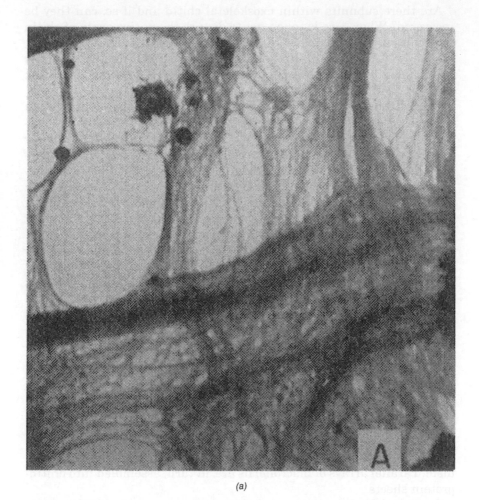

(a)

Figure 1. Electron microscopy of chitin assemblies. (a) "Chitin microfibers – presumably chitin micelles" prepared from large tracheae of a cockroach, *Periplaneta americana*. Preparation involved soaking in 5% NaOH at 20°C for 9 days, then washing in alcohol (magnification not given). (From Reference [1], p. 21). (b) An "electron micrograph of oblique section of chela tendon endocuticle of a crab, *Carcinus maenas*, showing helicoidal arrangement of microfibrils. The macrofibrils consist of parallel bundles of chitin crystallites and are seen in longitudinal and transverse section, with intermediate areas in oblique section. Decalcified in EGTA. (×38,000)." (Reproduced with permission from Reference [7a].)

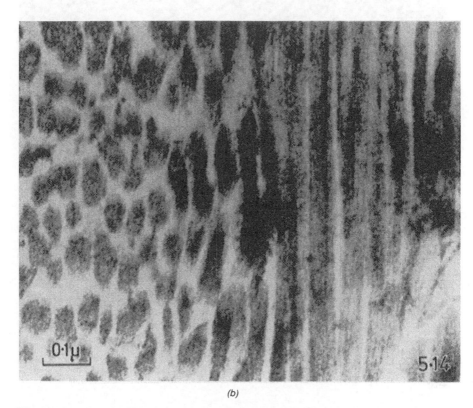

(b)

Figure 1 (continued). Electron microscopy of chitin assemblies. (b) An "electron micrograph of oblique section of chela tendon endocuticle of a crab, *Carcinus maenas,* showing helicoidal arrangement of microfibrils. The macrofibrils consist of parallel bundles of chitin crystallites and are seen in longitudinal and transverse section, with intermediate areas in oblique section. Decalcified in EGTA. (×38,000)." (Reproduced with permission from Reference [7a].)

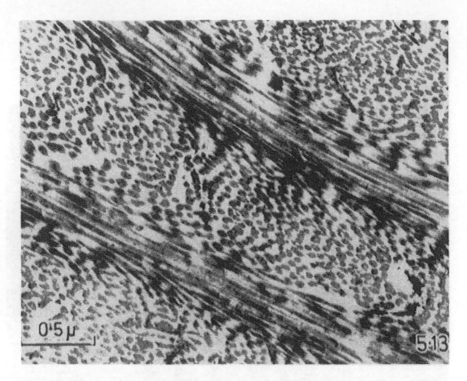

Figure 2. Electron microscopy of chitin assemblies II. Electron micrograph as in Figure 1(b), but at much higher magnification to show details of macrofibrils (× 120,000). (Reproduced with permission from Reference [7a].)

imately 10° rotated, until they again appear parallel at 180°, 360°, etc.

CHITIN: MAINTAINANCE OF STRUCTURE DURING PURIFICATION

When we began to inquire into the activity of chitin-specific enzymes in the molt of *Manduca sexta,* we discovered that a substrate was lacking that chitinases could readily recognize. This was evident from the dismal and irreproducible activity of chitinases against "purified chitin," "swollen chitin," "colloidal chitin," "decrease in viscosity of chitosan," etc., reported in the literature [10] and initially also found in my laboratory when commercial chitin was used as substrate. It was known (H. Lipke, personal communication) that chitinases cannot attack intact cuticle. Therefore, to assay chitinases, everyone thought it was necessary to isolate "purified" chitin, but the means for doing that had been rather mindlessly transferred from the wood treatment field

[11] to cuticles and carapaces, without considering the consequences on native chitin structure brought about by the chemical and mechanical means employed for isolation.

In the study of enzyme activity, intact native structure of the substrate is an obvious prerequisite if one wishes to obtain kinetically valid data. Bade and Wyatt [12] had described rapid transfer of carbon during the larval-to-pupal molt of what was then called *Platysamia cecropia.* The ^{14}C used to label the chitin in the cuticle of the mature caterpillar appeared as chitin of the same specific activity in pupal cuticle, whereas the surplus of chitin breakdown products from the last larval molt appeared in the pupal glycogen stores. Calculations showed that this massive Saturniid transfers up to 120 mg chitin to pupal stores in 12 h or less. Poor performance of molting fluid chitinases acting on "chitin" purchased from a scientific supply house demonstrated that either the substrate or else the moths originally observed must be wrong. One felt confident, however, that if a properly structured chitin was presented to these enzymes, rapid generation of product linearly with time could be expected.

Construction of a chitin substrate that would give these data was therefore pursued. Detailed inquiry into methods of chitin purification revealed that when chitinous animal structures are deproteinized with boiling NaOH (0.5–1 M), a form of chitin is obtained in which ultrastructure is maintained in latent form. This is shown in Figure 3. Use of hot, dilute NaOH (0.5–1.0 N, 100°C) leads to removal of all proteins from a number of arthropod cuticles and carapaces, resulting in highly purified Compacted Chitin [13]. It can be demonstrated that the native chain and fibrous structures are intact and stabilized in Compacted Chitin, presumably by mutual support of the intact chitin subassemblies, but the internal spacing that these subassemblies exhibit *in vivo* is absent [compare Figure 1(b) to Figure 3].

The method for insect cuticle deproteinization is applicable to calcified carapaces, provided the calcium salts are removed as a first step. Figure 4 shows what happens to native chitin structure when deproteinization is followed by exposure to a nonester-forming acid like HCl, which is customarily used in decalcification [14]. This "collapsed" form of chitin used to be typical of commercial chitins, which were produced by hot-base deproteinization of crab or shrimp carapaces followed by decalcification with HCl. Native structure cannot be reconstituted from such material because it is held in the collapsed form by the same multitude of hydrogen bonds that link sugar monomers in three dimensions in the native form [15]. Any exposure to nonester-forming acid of unprotected chitin will lead to collapse, even if, as in insects or horseshoe crabs, the exoskeleton was not originally calcified.

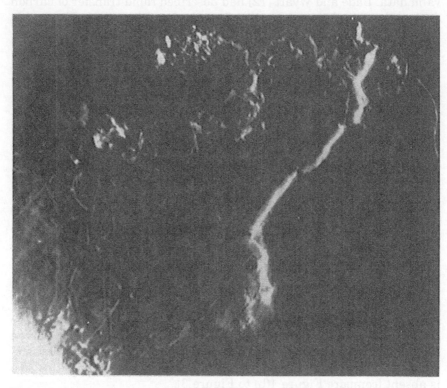

Figure 3. Light microscopy of chitin. A piece of Compacted shrimp Chitin, decalcified in RT 1 N HCl, deproteinized in boiling 0.5 M aqueous NaOH, ground in a Wiley micromill to 20 mesh, treated 10 min in 1 M KMnO$_4$, washed, and blenderized for 5–10 sec. Note Compacted Chitin macrofibrils and strands ravelled off (400×, Hofmann Modulation Optics).

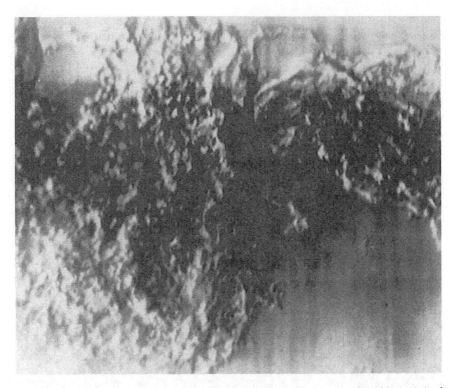

Figure 4. "Collapsed" chitin. Chitin treated in incorrect sequence, i.e., first deproteinized with NaOH, then treated with cool 1 N HCl (400×, Hofmann Modulation Optics).

Another way to damage chitin structure was found to be even brief exposure to bleaching agents. This effect also is seen in Figure 3. Fibers, similar to those shown, e.g., in Figure 1(a), and subject to attack by bacterial exochitinase, are seen to be split off by blenderizing a 20 mesh piece of Compacted shrimp Chitin following brief treatment with $KMnO_4$. H_2O_2, $KMnO_4$, ozone, and other oxidizing agents used in bleaching are equally deleterious to molecular structure of chitin: Compacted insect Chitin can be crumbled between one's fingers after exposure to H_2O_2.

It was eventually discovered that if properly prepared Compacted Chitin was allowed to disperse in an ice-cold 50–70% aqueous sulfuric or phosphoric acid solution, the following sequence of events ensued: An originally opaque mix of Compacted Chitin particles stirred in water would become transparent upon further addition of sulfuric acid and exhibit a Tyndall effect, while the viscosity rose sharply. When such a colloidal solution was then quenched in an excess of ice-cold water-ethanol (1:1, v/v), milk-white particles formed, which in the light microscope were strands rather than disorderly clumps. Our first glimpse of Linear Chitin in the light microscope is shown in Figure 5. If the acid-water mixture is not efficiently chilled during dispersal, irreversibly collapsed chitin is again produced, and a seemingly unbreakable gel forms if insufficient water is present in the reaction mixture during formation of the colloidal solution.

Freeze-drying strands of chitin formed under correct conditions gives a white, friable powder that is termed Linear Chitin. This material proved to be the sought-for substrate for chitin-specific enzymes. Chapter 3 in this book described studies on chitinase kinetics conducted with Linear Chitin and some other forms of chitin. Chitins of different biological origins prepared in this manner retain an architecture that is characteristic of the species: Lobster chitin with short and fat strands is of very different microscopic appearance than long, skinny shrimp chitin [Figure 6(a), (b)], and insect chitin, crab chitin, and

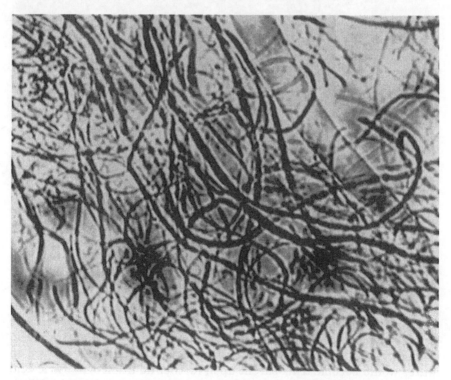

Figure 5. "Linear Chitin." Transmission light microscopy of *Manduca* chitin, SO$_4$-stabilized (400×).

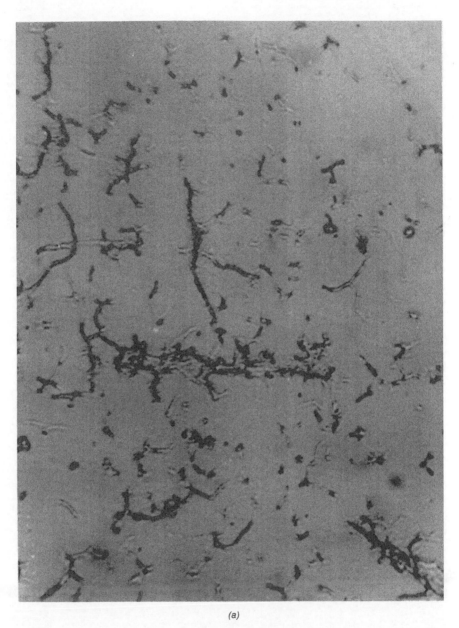

(a)

Figure 6. Linear Chitin from calcium-bearing organisms. (a) Linear Chitin from lobster carapace, *Homarus americanus* (400×, Hofmann Modulation Optics). (b) Linear Chitin from shrimp (400×, Hofmann Modulation Optics).

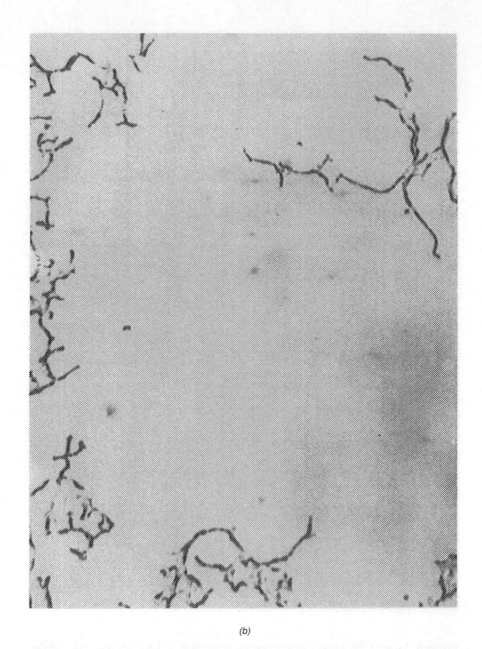

(b)

Figure 6 (continued). Linear Chitin from calcium-bearing organisms. (b) Linear Chitin from shrimp (400×, Hofmann Modulation Optics).

horseshoe crab chitin all exhibit reproducibly different configurations from lobster or shrimp chitins. In other words, although chemical composition appears to be similar or identical, Linear Chitins prepared from different biological sources exhibit striking differences in architecture. This finding confirmed what a number of authors [16] had surmised: Chitins of different biological origin seem to differ in efficacy in particular applications. Structural differences were also confirmed by differential rates of identical enzymes acting on Linear Chitins from various species (Bade, manuscript in preparation).

Unlike Compacted Chitin, Linear Chitin is of very great volume compared with its mass. Figure 7 demonstrates this. Volumes of 32-mg quantities of various chitins and chitin derivatives are compared here. Clearly, Linear Chitin occupies a far larger volume than all other prep-

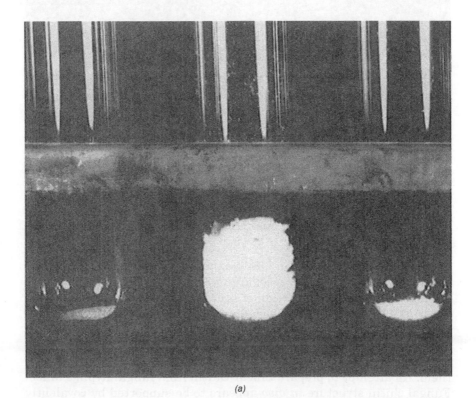

(a)

Figure 7. Volume of various forms of chitin. 32-mg quantities of a variety of chitin and chitin derivatives are shown. The tubes show left to right (a) "Microcrystalline" chitin; Linear Chitin (from *Manduca*); Compacted Chitin (*Manduca,* 20 mesh. (b) Japanese chitin (100 mesh, presumably Japanese Red Crab); commercial chitin from a domestic scientific supply house; (Japanese) carboxymethylchitin.

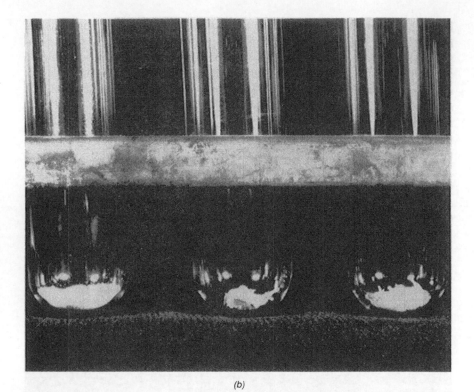

(b)

Figure 7 (continued). Volume of various forms of chitin. 32-mg quantities of a variety of chitin and chitin derivatives are shown. The tubes show left to right (a) "Microcrystalline" chitin; Linear Chitin (from *Manduca*); Compacted Chitin (*Manduca*, 20 mesh. (b) Japanese chitin (100 mesh, presumably Japanese Red Crab); commercial chitin from a domestic scientific supply house; (Japanese) carboxymethylchitin.

arations. Since Linear Chitin is also an active enzyme substrate giving kinetically valid data with specific enzymes, the case for the retention of native structure in formation of Linear Chitin is strengthened.

Earlier, Revol [17] published a photomicrograph showing short, stubby fibers decorating strands of Compacted Chitin that had been subjected to ultrasonic treatment. Evidently, then, there are chitin subassemblies, and it is possible to purify (animal) chitin to preserve them. An attempt to prepare Linear Chitin from fungal hyphae failed. Fungal chitin structure *in vivo* appears to be supported by covalently linked carbohydrates [18] rather than proteins as in animal chitins, and unlike the protein matrix surrounding animal chitin fibrils, which are rather readily removed, not all of the fungal glucans are subject to removal by hot base. It would be of interest to pursue this further. Since

Karrer, who maintained his views rather vociferously [19], conventional wisdom has held that all chitins are alike, but in the light of the observations cited this is clearly not the case.

CHITIN: OPTICAL CONFIRMATION OF STRUCTURE

A test for intact ultrastructure is birefringence. As early as 1907, Biedermann demonstrated birefringence in intact cockroach cuticle [20]. It is odd, therefore, that few employed an optical test for integrity of chitin structure during isolation, until retention of cuticle chitin birefringence in Compacted and in Linear Chitin was demonstrated by Bade [21]. This is seen in Figures 8(a–d). Figure 8(a) shows Linear Chitin in the polarizing microscope. Figures 8(b) and 8(c) are photomicrographs also taken under plane polarized light conditions. The samples analyzed in Figures 8(b) and 8(c) clearly are identical, but Figure 8(b) is Compacted Chitin, and Figure 8(c) shows Japanese commercial chitin. Figure 8(d), where little birefringence is demonstrable,

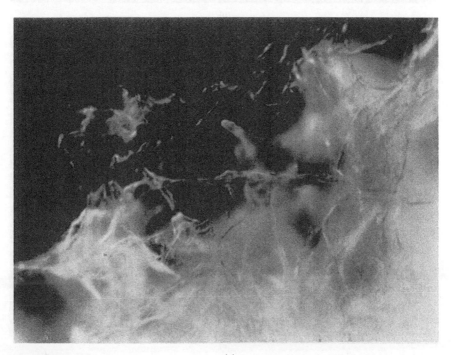

(a)

Figure 8. Chitin in plane polarized light. (a) Linear Chitin, (b) Compacted Chitin, (c) Japanese commercial chitin, (d) U.S. commercial chitin (all at 250× magnification).

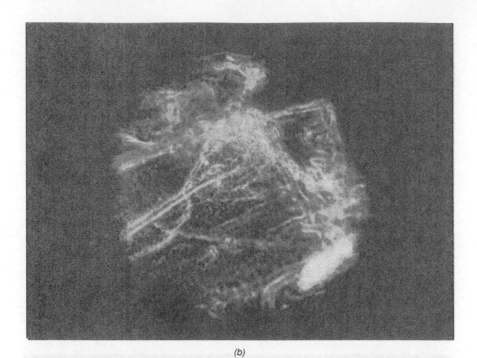

(b)

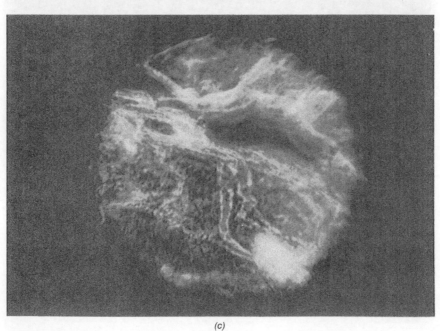

(c)

Figure 8 (continued). Chitin in plane polarized light. (a) Linear Chitin, (b) Compacted Chitin, (c) Japanese commercial chitin, (d) U.S. commercial chitin (all at 250× magnification).

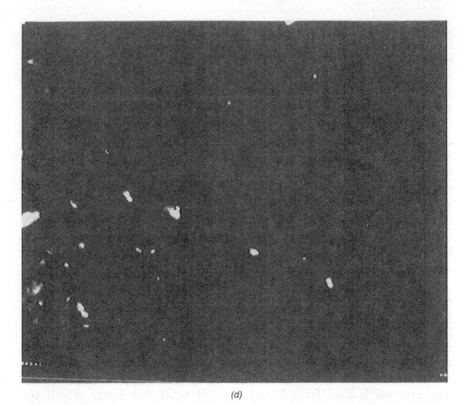

(d)

Figure 8 (continued). Chitin in plane polarized light. (d) U.S. commercial chitin (all at 250× magnification).

is a sample of commercial chitin purchased some years ago from a scientific supply house. Clearly, ultrastructure from the original chitinous structure can be retained through purification steps when animal chitin is processed correctly from carapace via Compacted Chitin to Linear Chitin. It is equally necessary to inquire into the history of preparation of a given chitin sample, whether obtained commercially or by a "standard" method practiced in the laboratory [14]. In other words, the structure purified chitin exhibits depends on the precise protocol for its isolation.

CHITIN: CHITIN-PROTEIN INTERACTION

What does the sequence of events described for obtaining linear chitin tell us about chitin structure *in vivo*? Proteins appear to protect the native structure of chitins from attack by chitinases and seemingly guard the chains from collapse of native structure under certain ambient conditions. When unprotected, i.e., totally deproteinized, chitin is exposed to acid, and if this acid is not ester forming, or an ester-forming acid at too high a temperature during formation of the colloidal chitin solution, the ultrastructure collapses when the subassemblies become separated from each other in sufficiently acid medium. From an observation quoted by A. G. Richards: "In soft cuticles [of insects], a large portion of the protein can be removed by water, especially hot water, while a smaller additional portion is removed by 5% KOH" [1]. Confirmed in many laboratories, one may conclude that although most of the cuticular protein is adsorbed to the chitin subassemblies and hence easily removed by hot water, some of it appears to be covalently linked to chitin, thus requiring base or protease for its removal. It is proposed that this covalently linked protein for which the name of chitogenin is here suggested, is the protein that maintains the chitin ultrastructure *in vivo*.

Several authors looked for linkage between free amine and aspartyl and histidyl residues in cuticle proteins, but that approach presupposes that the covalent linkage fortifying the native chitin structure is through an amide bond anchored at the nonacetylated nitrogen of glucosamine linked in the primary sequence. A consequence of this would be that chitin recovered after complete protein removal should be somewhat basic; the author is unaware that this type of experiment was ever conducted. Bade found that for total hydrolysis of sulfate-stabilized Linear Chitin, boiling with 6N HCl was required, whereas a bond linking sulfate and a primary amine group should be hydrolyzed by a much lower acid strength [22]. This hydrolysis was conducted on Linear Chitin stabilized with ^{35}S-labeled sulfate, and radioactive counts were extensively quenched until after total hydrolysis was achieved (Bade, manuscript in preparation). The ester bonds therefore must form internal cross-links in Linear Chitin, thus accounting for the much greater volume:mass ratio for Linear Chitin than for other chitin preparations (c.f. Figure 7). Since these ester functions are also nondialyzable and cannot be titrated, one concludes that they are covalently linked into Linear Chitin. Consider too that these ester cross-links form "instantly" under what would appear at first blush to be nonester-forming conditions. A hypothesis that unifies these various observations, and others, is that the cross-linking protein(s), as well as the cross-linking esters, are linked not to glucosamine or acetylated glucosamine, but to another highly reactive constituent forming part of the primary sequence. Linker proteins are known from other complex connective tissue assemblies, e.g., cartilage, and in particular from the cytoplasmic component glycogen where a number of peptides are associated with initiation and elongation [23].

If a constituent suitable for very rapid esterification exists within the primary sequence, it should be much more highly reactive than N-acetylglucosamine, and this hypothetical constituent may not even be a sugar. When esterifiable sites on N-acetylglucosamine are considered, there are two free OH groups, one each on C3 and C6, which suggests that every twentieth to twenty-fifth residue is esterified; an amide linkage to SO_4 has been ruled out by the hydrolysis data cited above. A rough calculation indicates that if one in fifty residues within the primary sequence of animal chitins were an entity other than N-acetylglucosamine, approximately 1% by weight would be obtained for crosslinking SO_4 groups, which is the ratio for the SO_4:N-acetylglucosamine masses measured for Linear Chitin. Such substitution of a non-N-containing entity at regular intervals along the chitin chains, more reactive and more easily destroyed than glucosamine, would also explain some other discrepancies met with in the chitin literature. For example, elementary analysis of "pure" chitins occasionally are reported to have a smaller percentage of N than expected [24]. One would look for the opposite if some unacetylated glucosamines were involved in the primary sequence to which chitogenin, or in Linear Chitin covalently linked esters, would be attached. Further, glucosamine yields from acid hydrolysis or methanolysis rarely exceed 70% of the amount calculated from the chitin mass being analyzed. There is, of course, a "fudge factor" argument that says that glucosamine is just so very labile that 30% is destroyed, but in the age of nano-biochemistry that is unsatisfactory. It should further be pointed out that the quantity of ester bonds formed, to Compacted Chitin reacted, during preparation of Linear Chitin is invariant within limits of experimental error. Finally, the hierarchy of chitin structures that can be demonstrated in cuticles and carapaces extends over a number of cell surfaces presented by the hypodermis, so that a "steering mechanism" for *in vivo* construction of these assemblies seems called for; the hypothetical linker chitogenin could play a role in "steering" during assembly, as well as in stabilization of ultrastructure, before its removal by hydrolysis in hot base. In any event, the existence of Linear Chitin now permits the necessary experiments for testing this notion because pure chitinases used sequentially will permit isolation of esters bearing the entity that serves in crosslinking, and hot water extraction of chitin-protein matrix without prior use of base should give a substrate for the exploration of the existence and identity of chitogenin by similar enzymatic means.

CHITIN: PARALLEL OR ANTIPARALLEL

Chitin chains have one reducing and one nonreducing end. The antiparallel arrangement of chains, i.e., ones where adjacent chains have reducing and nonreducing ends juxtaposed, has been proposed by X-ray crystallographers, whereas others have proposed a parallel structure [7] (cf. Chapter 3 in this book). Note that the "typical α-structure" scatter diagram is obtained "by boiling [e.g., lobster tendon] in 5% potassium hydroxide" (which one would expect to lead to extensive deacetylation – MLB), ". . . soaked in cold 3 N hydrochloric acid for 3 hrs . . . and subjected to a further treatment with 6 N HCl for 10 minutes to increase the crystalline order. . . . The specimen of α-chitin used to obtain the x-ray data was in form of a thin sheet peeled from the purified tendon" [15]. Or: "The sample for diffraction photography was prepared by rolling in the shape of a needle, a thin sheet of chitin [from the common black Indian scorpion]. The chitin sheet had previously been purified by treating the cuticle in 5% cold dilute KOH for about 7 days and then with dil. HCl for about half an hour" [6]. In view of the profound effect on linear structure that a nonester-forming acid like HCl can have and the fact that in lobster tendon, the chains are constrained in a lengthwise arrangement, variations *in vivo* of the antiparallel unit cell of chitin are conceivable. Possible changes in chitin structure have been discussed by Rudall [25].

CHITIN: SUMMARY

It should be apparent that possession of Linear Chitin now opens to rational inquiry a number of outstanding questions regarding chitin. Isolation of ester- and/or protein-linked components in the primary sequence by use of precisely defined (and probably gene-derived) chitinases is one. Also now accessible are an inquiry into the precise structure of chitins exhibiting different X-ray patterns, whether α-chitin construction is parallel as opposed to antiparallel; precise knowledge of the amount of reactive substrate present in enzyme reactions, as for instance deacetylases; and a realistic hope that action of deacetylase(s) will produce chitosans with a reproducible pattern of deacetylation. This is so because Eveleigh et al. [26] demonstrated that chitosanases are dependent on a specific chitosan structure, viz., one varying between 95% and 65% acetylation, and were inactive outside this region. This implies a reproducible, rather than a haphazard, pattern of deacetylation, and since in nature chitosan is derived enzymatically from chitin [27], deacetylase(s) may be expected to work to a reproducible pattern.

CONCLUSIONS

1. Animal chitin *in vivo* has highly ordered molecular structure. This

is demonstrated by rapid chitin recycling in the arthropod molt due to enzyme activity.

2. Such chitin can be purified to maintain the highly ordered structure it possesses *in situ.*
3. Both retention of native ultrastructure and recapture of spacing between chitin subassemblies are required to give kinetically valid data in the study of specific enzymes, e.g., chitinases and chitin deacetylases.
4. Other outstanding problems regarding chitin structure and performance can now be successfully studied.

ACKNOWLEDGEMENTS

Previously, this work was supported by National Institutes of Health and National Science Foundation, before work on environmental matters became unfashionable. Grateful acknowledgement is made to my co-workers: Alfred Stinson, Senior Research Associate; Michael Kosmo, J. J. Shoukimas, Ivan Boyer, Kerry Hickey, and L. Lapierre, students; Nehad A.-M. Moneam, postdoctoral associate, and G. R. Wyatt, Ph.D. thesis advisor.

REFERENCES

1. Richards, A. G. 1951. *The Integument of Arthropods.* Minneapolis: University of Minnesota Press.
2. Braconnot, H. 1811. "Récherches Analytique sur la Nature des Champignons." *Ann. Chim.,* 79:265–304.
3. Odier, A. 1823. "Mémoire sur la Composition Chimique des Parties Cornées des Insectes." *Mém. Soc. Hist. Nat. Nat. Paris,* 1:29–42.
4. Ledderhose, G. 1876. "Über Salzsäures Glycosamin." *Ber. Dtsch. Chem. Ges.,* 9:1200–1201.——1878/79. "Über Chitin und Seine Spaltungsprodukte." *Z. Physiol. Chem.,* 2:213–227.
5. Levene, P. A. 1921. "Synthese von 2-Hexosaminsäuren und Hexosaminen." *Biochem. Z.,* 124:37–83.
6. Sundara Rajulu, G. and Gowri, N. 1978. "Chitin from Marine Organisms and Its Use as an Adhesive." *Proc. 1st Internat. Confer. Chitin/Chitosan* (R. A. A. Muzzarelli and E. R. Pariser, eds.), Cambridge: MIT Press, pp. 430–436.
7. Rudall, K. M. and Kenchington, W. 1973. "The Chitin System." *Biol. Rev.,* 49:597–636.
7a. Neville, A. C. 1975. *Biology of the Arthropod Cuticle.* New York: Springer-Verlag.
8. Muzzarelli, R. A. A. 1977. *Chitin.* Elmsford, NY: Pergamon Press, pp. 52–55.

8a. Ogawa, K., Okamura, K. and Hirano, S. 1982. "Ultrastructure of Chitin in the Cuticles of a Crab Shell Examined by SE Microscopy and X-ray Diffraction." In *Chitin and Chitosan.* 2nd Confer. (S. Hirano and S. Tokura, eds.) Tottori, Japan, pp. 87–92.

9. Bouligand, Y. 1972. "Twisted Fibrous Arrangements in Biological Materials and Cholesteric Mesophases." *Tissue Cell.,* 4:189–217.

10. Chitinase Assays. 1988. Sect. [48]–[50]. In: *Methods in Enzymology, Vol. 161: Biomass Pt. B: Lignin, Pectin, and Chitin.* (W. A. Wood and Scott T. Kellogg, eds.) San Diego, CA: Acad. Press; also spread throughout the chitin literature.

11. Wood, T. M. and J. N. Saddler. 1988. "Increasing the Availability of Cellulose in Biomass Material." In: *Meth. Enzymol. 160 Biomass Pt. A: Cellulose and Hemicellulose,* (W. A. Wood and Scott T. Kellogg, eds.), San Diego: Acad. Press, pp. 1–11.

12. Bade, M. L. and G. R. Wyatt. 1962. "Metabolic Conversions during Pupation of the Cecropia Silkworm. I: Deposition and Utilization of Nutrient Reserves." *Biochem. J.,* 83:470–478.

13. Bade, M. L., A. Stinson and Nehad A.-M. Moneam. 1988. "Chitin Structure and Chitinase Activity: Isolation of Intact Chitins." *Connect. Tissue Res.,* 17:137–151.

14. Muzzarelli, R. A. A. 1977. *Chitin.,* Elmsford, NY: Pergamon Press, pp. 89–94.

15. Blackwell, J., R. Minke and K. H. Gardner. 1978. "Determination of the Structures of α- and β-Chitins by X-ray Diffraction." *1st Internat. Confer. Chitin/Chitosan* (R. A. A. Muzzarelli and E. R. Pariser, eds.), Cambridge, MA: MIT Sea Grant Program, pp. 108–124..

16. Balassa, L. L. and J. F. Prudden. 1978. "Applications of Chitin and Chitosan in Wound-healing Acceleration." *Proceed. 1st Internat. Confer. Chitin/Chitosan,* (Muzzarelli, R. A. A., and E. R. Pariser, eds.), pp. 296–305. Also: Averbach, B. L. 1978. ibid. "Film-forming Capacity of Chitosan," Cambridge, MA: MIT Sea Grant Program, pp. 199–209.

17. Revol, J. F. 1989. "Lattice Resolution in α-Chitin," *Internat. J. Biol. Macromol.,* 11:233–235.

18. Kollàr, R., E. Pétraková, G. Ashwell, P. W. Robbins and E. Cabib. 1995. "Architecture of the Yeast Cell Wall. The Linkage between Chitin and $\beta(1 \rightarrow 3)$-Glucan," *J. Biol. Chem.,* 270:1170–1178.

19. Karrer, P. and A. Hoffmann. 1929. "Polysaccharide XXXIX. Über den Enzymatischen Abbau von Chitin und Chitosan," *Helv. Chim. Acta,* 12:616–637; Karrer, P. 1930. "Der Enzymatische Abbau von Nativer und Umgefällter Zellulose, von Kunstseiden und von Chitin," *Kolloid. Z.,* 52:304–319.

20. Biedermann, W. 1907. "Über die Struktur des Chitins bei Insekten und Crustaceen," *Anat. Anz.,* 21:485–490; [1], pp. 130–132.

21. Bade, M. L. 1993. "Separations and Reactions Based on Superior Forms of Chitin and Chitosan." Ch. 18 in: *Cellulosics: Materials for Selective Separa-*

tions and Other Technologies. (J. F. Kennedy, G. O. Phillips and P. A. Williams, eds.) London: Ellis Horwood, pp. 155–162.

22. Whistler, R. L. and W. W. Spencer. 1963. "Sulfation." In: *Methods in Carbohydrate Chemistry. Vol. III* (R. L. Whistler, ed.), New York: Acad. Press, pp. 265–266.

23. Hart, G. W., R. S. Haltiwanger, G. D. Holt and W. G. Kelly. 1989. "Glycosylation in the Nucleus and Cytoplasm: Glycogenin." In *Ann. Rev. Biochem.,* 58:841–874.

24. BeMiller, J. N. 1965. "Chitin." In: *Meth. Carbohydr. Chem.* (R. L. Whistler, ed.) Vol. V, pp. 103–105.

25. Rudall, K. M. 1976. "Molecular Structure in Arthropod Cuticles." Ch. 2 in: *The Insect Integument* (H. R. Hepburn, ed.), Amsterdam: Elevier Scientific Publ. Co., pp. 21–41.

26. Davis, B. and D. E. Eveleigh. 1984. "Chitosanases: Occurrence, Production, and Immobilization." in: *Chitin, Chitosan, and Related Enzymes* (J. P. Zikakis, ed.), Orlando, FL: Acad. Press, pp. 161–180.

27. Akari, Y. and E. Ito. 1975. "Pathway of Chitosan Formation in *Mucor rouxii.*" *Eur. J. Biochem.,* 55:71–78.

PART V

TEXTILES AND POLYMERS

CHAPTER 16

Fine Structure and Properties of Filaments Prepared from Chitin Derivatives

G. W. URBANCZYK

Institute of Fiber Physics and Textile Finishing
Technical University of Lodz
Lodz, Poland

INTRODUCTION

Chitin and their derivatives such as chitosan and butyrylchitin reveal very distinctly pronounced bioactivity, hemocompatibility, and antithrombotic properties [1–6]. These polymers are therefore utilized with good effects as special products in surgical treatment of the human body.

In contrast to chitin, chitosan and butyrylchitin are characterized by a good fiber-forming ability [5–7]. Filaments spun from chitosan and butyrylchitin can be employed in the manufacture of woven and knitted textile goods utilized in medicinal dressing materials, as an exclusive or additional fibrous component.

The application of chitosan and butyrylchitin filaments in the above areas is however still limited. The main reason is insufficient knowledge about the fine structure and the basic physical and chemical properties of chitin derivatives. The results of investigations presented in this chapter should bring about an extension of the knowledge about fibers made of chitosan and butyrylchitin.

EXPERIMENTAL

Materials

Two kinds of filaments were studied: chitosan filaments of Japanese

281

and Polish origin, and dry- and wet-spun butyrylchitin filaments with differentiated polymer molecular weight. In the case of the Japanese chitosan filaments, they were spun using the wet spinning scheme from a 15% solution of chitosan in acetic acid into an alkaline coagulation bath. The chitosan was characterized by a deacetylation degree of 61% and molecular weight of 40,000. Polish chitosan filaments were spun also with a wet spinning scheme from 2.45% solution of chitosan in 4% acetic acid into a coagulation bath containing 5% NaOH and 40% CH$_3$OH at 30°C. The utilized chitosan was characterized by a deacetylation degree of 62.8% and a molecular weight of 444,600. Both kinds of chitosan filaments were drawn in the final stage of preparation to 20% extension.

The butyrylchitin filaments were manufactured as follows: dry-spun fibers were obtained by spinning from a 23% solution of butyrylchitin in acetone into air. Three kinds of filament were prepared; low molecular weight ($\eta = 1.23$), medium molecular weight ($\eta = 1.28$), and high molecular weight ($\eta = 1.44$) butyrylchitin. The wet spinning was performed from a 16% solution of high molecular weight butyrylchitin ($\eta = 1.44$) in dimethyleformamide into a water coagulation bath.

RESULTS

Macrostructure

The filaments exhibit a markedly pronounced differentiation in the shape of their cross sections. The pictures of the relevent cross sections are presented in Figures 1, 2, and 3. As can be seen, chitosan and wet-spun butyrylchitin filaments are characterized by a relatively regular, nearly circular shape in their cross sections. In contrast, the cross sections of dry-spun butyrylchitin filaments were very irregular, elongated, and showed a long contour line.

The results of the quantitative appraisal of the cross sections assessed by means of the average values of the area (S) and the coefficient of shape development (W) are presented in Table 1. The measurements of the optical density across the filament cross section revealed an even intensity distribution which is evidence of the absence of skin-core building in all the examined filaments. Further, on examination of the filaments under a polarization microscope (magnification 600×), did not reveal maltese cross brightening patterns. This enabled us to conclude that sperulite crystalline aggregations were not present in the filaments.

Figure 1. Cross sections of Japanese chitosan filaments.

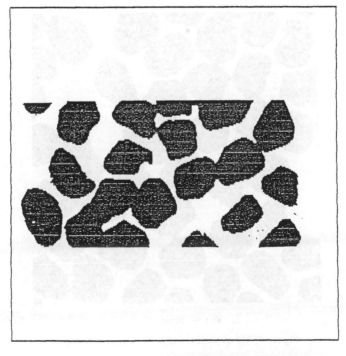

Figure 2. Cross sections of Polish chitosan filaments.

Dry - Spun

Wet - Spun

Figure 3. Cross sections of butyrylchitin filament.

Table 1. Filament cross section parameters.

Parameter	Chitosan Filament		Butyrylchitin Filament	
	Japanese	Polish	Dry-Spun	Wet-Spun
Mean cross section Area—S (μm^2)	289 $V_S = 7.1\%$	90 $V_S = 11.9\%$	1196–1931	439
Mean length of the contour line— L (μm)	84.1	87.9	215.1–238.1	83.9
Mean value of the shape developing coeff.—W_k	0.065	0.133	0.541–0.755	0.130

V: coefficient of variation.
W_k: for circle shape = 0.000.

Microstructure

Crystalline Structure

X-ray diffraction examination of the filaments showed a lattice organization. The lattice corresponded to the orthorhombic crystallographic system. However, it should be emphasized that there was a difference in the lattice unit cells between chitosan and butyrylchitin filaments. This discrepancy refers to the b-dimension of the unit cells. The volume of the unit cell and the theoretical crystalline density were also different. It should be added that the lattice in chitosan filaments refers to the II-type crystalline structure of chitosan based on the description of Samuels [8]. This last peculiarity was proved not only by X-ray diffraction but also by IR spectroscopy. The IR absorption spectrograms showed absorption bands at 1,670 cm^{-1}, 1,425 cm^{-1}, and 1,380 cm^{-1} characteristic for the II-type crystalline chitosan. Absorption bands at 760 cm^{-1} and 1,350 cm^{-1}, typical for the I-type crystalline chitosan, were not observed. The basic findings of the lattice organization in the crystalline regions of the investigated filaments are shown in Table 2.

The data obtained from X-ray diffraction investigations pertaining to the degree of crystallinity and the average lateral crystallite size perpendicular to the lattice planes (100) $- D_{100}$ are presented in Table 3. As one can see, the chitosan and butyrylchitin filaments differ considerably in their crystallinity degree and the average lateral crystallite size. The degree of crystallinity for chitosan filaments is about 0.37–0.38, whereas for butyrylchitin filaments it is 0.17–0.19. The average lateral crystallite sizes differ very little and are respectively

Table 2. Lattice parameters of crystalline regions.

Kind of Filament	Unit Cell Dimensions			Volume of Unit Cell ($Å^3$)	Theoretical Cryst. Density (g/cm^3)
	a (Å)	b (Å)	c (Å)		
Chitosan (Japanese and Polish)	4.4	10.0	10.3	133.6	1.7698
Butyrylchitin (dry- and wet-spun)	4.4	13.4	10.3	292.8	1.8758

20.6–21.5 Å and 22–23 Å. The values of the crystallinity degree provide evidence that because of the differences in the molecular structure of both polymers (i.e., longer side groups of the butyrylchitin chains) chitosan is characterized by a greater crystallization ability. The nearly same lateral dimensions of crystallites and the essential differences in the degree of crystallinity allow us to conclude that the number of crystalline regions in chitosan filaments is bigger than in butyrylchitin filaments.

Internal Orientation

From the data collected, Table 4, the filaments also show markedly pronounced differences with respect to the degree of internal orientation. Chitosan filaments, and especially Japanese compared to butyrylchitin, reveal a better internal orientation. The overall orientation evaluated by means of birefringence is about 10 times larger than butyrylchitin filaments.

The crystalline orientation is similarly better. The value of the crystallite orientation factor f_x is twice as large as for butyrylchitin filaments. The X-ray diffraction studies of crystallite orientation provide additional evidence that the orientation of crystallites in both

Table 3. Crystalline structure parameters.

Kind of Filament	X-ray Degree of Crystallinity	Average Lateral Crystallite Size—$D_{(100)}$ (Å)
Chitosan Japanese	0.37	20.6
Chitosan Polish	0.38	21.5
Butyrylchitin dry-spun	0.17–0.19	22.0–23.0
Butyrylchitin wet-spun	0.17	22.0

Table 4. Internal orientation parameters.

Kind of Filament	Birefringence	Crystallite Orientation Factor	Plate Effect Coefficient	Dichroic Ratio of Absorption Band $R = D_{\parallel}/D_{\perp}$						
				730 cm^{-1}	941 cm^{-1}	1670 cm^{-1}	661 cm^{-1}	893 cm^{-1}	1260 cm^{-1}	1654 cm^{-1}
Chitosan Japanese	+0.017	0.6656	1.45	–	–	–	0.515	0.538	0.656	0.879
Chitosan Polish	+0.014	0.6389	1.39	–	–	–	0.565	0.842	0.724	0.841
Butyrylchtin dry-spun	none	0.3139–0.3235	none	1.407–1.581	1.100–1.415	0.987–1.092				
Butyrylchitin wet-spun	–0.0016	0.3169	none	0.988	0.338	0.853				

kinds of filaments differs. Qualitatively the chitosan filaments manifest themselves in a well-pronounced, so-called orientation plate effect, whereas for butyrylchitin filaments such an effect does not appear. This in turn enables us to conclude indirectly that the crystalline phase of chitosan filaments consists of plate-like, strongly mutually joined crystallites; whereas in the case of butyrylchitin filaments, they consist of rod-like crystallites only weakly joined to each other.

Physical Properties

Mechanical Properties

Parameters characterizing the mechanical properties are presented in Table 5. Analyzing the values one can see that common to both types of filaments but especially for butyrylchitin filaments the tensile strength is very small. From this point of view, they can be included in the group of weakest man-made fibers made of natural polymers.

The filaments also differ in respect to their extensibility, qualitatively as well as quantitatively. The stress-strain curves are presented in Figure 4. The following conclusions can be drawn. A slightly viscoelastic extensibility refers to the Polish chitosan filaments and is better expressed in the case of wet-spun butyrylchitin filaments. On the other hand, Japanese chitosan, and even more so, the dry-spun butyrylchitin filaments reveal a distinctly viscoelastic deformability. These conclusions are corroborated by the established values of elongation at break. They are very small for the Polish chitosan and wet-spun butyrylchitin filaments (5.6% and 8.3%) and relatively larger for Japanese chitosan and dry-spun butyrylchitin filament (14.8% and 33.9–47.8%).

Table 5. Mechanical properties at RH = 65% and T = 20°C.

Kind of Filament	Breaking Load F F = (N)	Tensile Strength σ (MPa)	Strain at Rupture ε (%)	Young's Modulus E (GPa)
Chitosan Japanese	0.0414 V = 7.01 (%)	134.3	14.8 V = 14.4 (%)	2.45 V = 18.1 (%)
Chitosan Polish	0.0140 V = 13.0 (%)	156.3	5.6 V = 22.8 (%)	5.56 V = 14.9 (%)
Butyrylchitin dry-spun	0.1168–0.2335 V = 12.3 (%)	88.8–120.9	33.9–47.8 V = 15.7 (%)	1.62–2.06 V = 18.1 (%)
Butyrylchitin wet-spun	0.0451 V = 18.7 (%)	102.7	8.3 V = 26.8 (%)	2.94 V = 15.4 (%)

V: coefficient of variation.

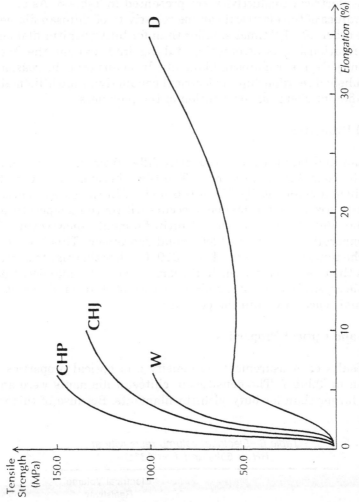

Figure 4. Stress-strain curves of single filament: CHP– Polish chitosan filaments, CHJ –Japanese chitosan filaments, W–wet-spun butyrylchitin filaments, and D–dry-spun butyrylchitin filaments.

Electrical Properties

Both types of filaments differ considerably in their electrical conductivity. The results of measured electrical volume resistivity characterizing the electrical conductivity are presented in Table 6. As one can see, the values of the electrical volume resistivity of chitosan filaments is on the order 10^2–10^4 times smaller than for butyrylchitin filaments. Such a relationship is astonishing, taking into account the higher crystallinity degree of chitosan filaments. It means that the reason for the established relationship in electrical conductivity must lie mainly in the different chemical constitution of the polymers.

Thermal Properties

Chitosan and butyrylchitin filaments differ from each other qualitatively in their thermal properties. The DSC thermograms of the filaments differ fundamentally (Figures 5 and 6). In the case of chitosan filaments a strong exothermal peak occurs within the temperature interval 300–350°C, whereas butyrylchitin filaments show within that same temperature range an endothermal depression. This is evidence that in the temperature interval 300–350°C other thermal transitions occur; in the case of butyrylchitin it seems to be a thermal decomposition of the crystalline regions, and in the case of chitosan filaments an exothermal reaction within the polymer.

Density and Optical Properties

The results of measurements of density and optical properties are presented in Table 7. The densities of chitosan filaments were about 15–18% larger than for butyrylchitin filaments. Because of this rela-

Table 6. Electrical volume resistivity at
RH = 65% and T = 23°C.

Kind of Filament	Electrical Volume Resistivity ϱ_V ($\Omega \cdot$cm)
Chitosan Japanese	$2.8 \cdot 10^7$
Chitosan Polish	$1.24 \cdot 10^5$
Butyrylchitin dry-spun	$6.5 \cdot 10^9$
Butyrylchitin wet-spun	$1.84 \cdot 10^9$

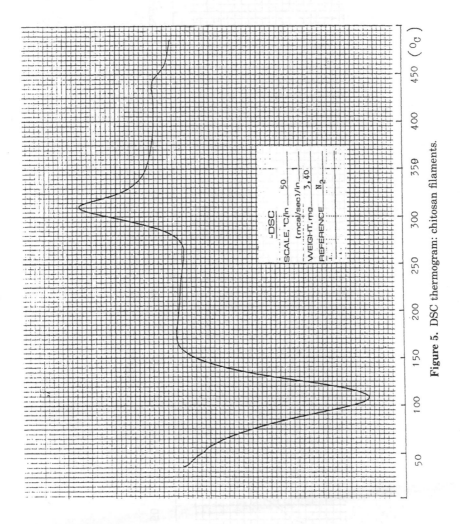

Figure 5. DSC thermogram: chitosan filaments.

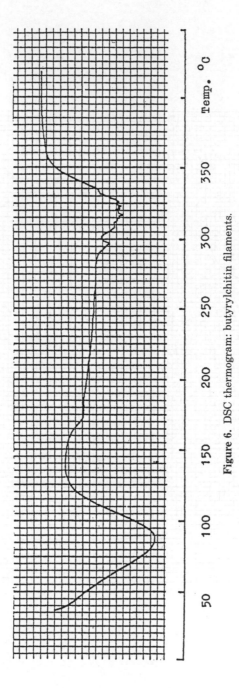

Figure 6. DSC thermogram: butyrylchitin filaments.

Table 7. Density and optical properties.

Kind of Filament	Density g/cm^3	Directional Refractive Index		Isomorphic Refractive Index
		$n_{\parallel}$	$n_{\perp}$	n_{iso}
Chitosan Japanese	1.452	1.5530	1.5357	1.5415
Chitosan Polish	1.427	1.5412	1.5270	1.5317
Butyrylchitin dry-spun	1.2370–1.2390	—	—	1.4923–1.4939
Butyrylchitin wet-spun	1.2439	1.4861	1.4876	1.4871

tionship the optical properties reveal a similar differentiation. The values of the isomorphic refractive index for chitosan filaments are markedly larger than for butyrylchitin filaments. The filaments also differ significantly in their optical character and the anisotropy of light transmission. Chitosan filaments are distinctly optically positive, whereas wet-spun butyrylchitin filaments are negative (negative birefringence). The optical anisotropy is strongest in the Japanese and in the second order in the Polish chitosan filaments. Very small anisotropy pertains to the wet-spun butyrylchitin filaments. In the case of dry-spun butyrylchitin the birefringence is not measurable by means of conventional measuring techniques. The filaments do not show any anisotropy.

Physicochemical Properties

Hygroscopicity and Swelling

The data characterizing the hygroscopicity of filaments and the swelling values in water and 0.9% water solution of physiological salt at 20°C are presented in Table 8. As one can see, both types of chitosan

Table 8. Normal humidity, swelling in water and 0.9% water solution of physiological salt at 20°C.

Kind of Filament	Normal Humidity (%)	Swelling in H$_2$O (%)	Swelling in Physiological Water Solution (%)
Chitosan Japanese	13.6	57.0	34.0
Chitosan Polish	18.1	75.0	61.8
Butyrylchitin	6.5	10.0	—

Table 9. The up-take of the dyestuffs.

| | Dyeing Parameters | | | Dyestuff Uptake (mg dye/100 mg fiber) | | Butyrylchitin | |
Class of the Dyestuff	Dyestuff Color Index Number	Temperature (°C)	Concentration of the Bath: % of the Fiber Weight	Chitosan (Japanese)	Chitosan (Polish)	Dry-Spun	Wet-Spun
Direct dyestuff	Direct Yellow 12/24895	100	2%	0.60	0.35	—	—
Acid dyestuff	Acid Red 1/18050	100	2%	0.34	0.24	—	—
Reactive dyestuff	Reactive Red 246/4896	100	2%	1.39	0.859	—	—
Disperse dyestuff	Disperse Yellow 3/11855	100	2%	0.008	0.010	0.162–0.173	0.168

filaments exhibit considerably larger values of normal humidity than butyrylchitin filaments. A similar relationship was found in swelling in water. The relationships were quite unexpected and amazing. Regarding that, the chitosan filaments revealed higher crystallinity and better internal orientation. It seems that such a layout of humidity and swelling must be the result of specific differences in the chemical structure of the polymers.

Dyeability

The preliminary qualitative assessment of dyeability provided evidence that chitosan filaments, irrespective of their origin, are susceptible to dyeing by all the main classes of dyestuffs. Butyrylchitin though, could only be dyed by means of disperse dyes. The dyeability of filaments has been evaluated quantitatively on the basis of dyestuff uptake. From the dyestuff uptake values presented in Table 9, it may be inferred that in the case of chitosan filaments the reactive dyestuffs are the most efficient, followed by the direct dyes. Acid and particularly disperse dyestuffs are markedly less effective at the same dyeing parameters. The results prove the very good dyeability of butyrylchitin filaments with respect to disperse dyes.

CONCLUSIONS

The studied filaments regarded as raw materials for manufacture of medical dressing materials have advantages but also some negative properties. They have relatively high hygroscopicity, considerably large swelling ability in water and physiological salt solution. In contrast, the low tensile strength and relatively low extensibility, with exception of the dry-spun butyrylchitin filaments, are drawbacks. These properties depend on the fine structure of the filaments. The first one is a result of the relatively small crystallinity. The low-tensile strength seems to be a consequence of the extremely low internal overall orientation of the filaments.

REFERENCES

1. W. C. Malette, H. J. Quingly, R. D. Gaines, N. D. Johnson and W. G. Rainer. 1983. *Ann. Thorac. Surg.*, 36:555.
2. S. Hirano and Y. Noishiki. 1983. *J. Biomed. Rev.*, 19:413.
3. J. Dutkiewicz, et al. 1983. *Polym. Med.*, 28:177.
4. G. C. Allan, L. C. Altman, R. E. Bensinger and D. K. Chosk. 1988. In *Programme and Abstracts of 4th Conference on Chitin/Chitosan,* Trondheim, Norway.

5. S. Grant, H. S. Blain and G. McKay. 1983. *Macrom. Chem.,* 190:2279.

6. J. Dutkiewicz, L. Szosland, M. Kucharska, C. Judkiewicz and R. Ciszewska. 1990. *J. Bioact. Compat. Polymers.* 5:293.

7. S. Tokura, S. Baba, Y. Uraki, Y. Miura, N. Nishi and O. Hasekawa. 1990. *Carbohydr. Polym.,* 13:273.

8. R. J. Samuels. 1981. *J. Pol. Sci. Phys. Ed.,* 19:1081.

CHAPTER 17

Graft Copolymers

KEISUKE KURITA
Department of Industrial Chemistry
Faculty of Engineering
Seikei University
Musashino-shi, Tokyo 180, Japan

INTRODUCTION

Chitin and chitosan are amino polysaccharides and thus anticipated to have high potential as specialty polymeric materials compared with cellulose. They are, however, utilized scarcely because of problems associated with poor solubility, multifunctionality, and low reactivity. To develop advanced functions leading to utilization, much attention has been paid to chemical modifications of chitin and chitosan. Of various potential modifications, graft copolymerization is one of the most attractive approaches to versatile molecular designs; it will afford novel types of tailored hybrid materials composed of natural polysaccharides and synthetic polymers. The properties of the graft copolymers would be widely controlled by the characteristics of the introduced side chains, including the molecular structure, length, and number.

Although graft copolymerization onto chitin and chitosan has started attracting attention only recently, currently progressing studies on the various modes of graft copolymerization and on the relationship between the molecular structure and properties will open a way to practical utilization of these significant biomass resources.

This chapter describes the procedures for graft copolymerization onto chitin and the derivatives as well as the properties of the resulting graft copolymers. The grafting behavior is often discussed in terms of grafting percentage and grafting efficiency defined as follows: grafting percentage = (weight of introduced polymer branches/weight of trunk polymer) × 100; grafting efficiency = (weight of introduced polymer branches/weight of monomer charged) × 100.

GRAFT COPOLYMERIZATION ONTO CHITIN

Copolymerization of Vinyl Monomers

Although graft copolymerization of vinyl monomers onto chitin is conducted by several initiating methods, those with initiators have been most extensively studied thus far. Cerium(IV) is sometimes used for grafting onto cellulose, and it is also effective for grafting onto chitin.

The reaction of powdery chitin suspended in water with acrylamide or acrylic acid proceeds efficiently in the presence of cerium at 60°C. The amount of cerium affects the polymerization markedly, and under appropriate conditions, the grafting percentage reaches about 200% (Kurita et al., 1991). The resulting graft copolymers show improved solubility and swelling; those having poly(sodium acrylate) side chains are soluble in dichloroacetic acid, whereas those having polyacrylamide chains swell highly. Both of the copolymers exhibit much enhanced hygroscopicity compared with chitin.

Methyl acrylate (Lagos et al., 1992) and methyl methacrylate (Ren et al., 1993) are similarly graft-copolymerized with cerium. Chitin-*graft*-poly(methyl methacrylate) swells in *N,N*-dimethylformamide (DMF), and the extent of swelling is dependent on the grafting percentage. When the grafting percentage is below 400%, the copolymer shows lower swelling than the original chitin. Above 400% grafting, however, the swelling becomes profound to give transparent gels, from which films are cast. The glass transition temperature of the copolymer is 130°C as determined by differential scanning calorimetry (Ren et al., 1993).

Grafting of methyl methacrylate onto chitin is also possible with tributylborane in water but not in organic solvents. Grafting efficiency is generally low but can be raised up to about 30% with prolonged reaction at 37°C (Kojima et al., 1979).

γ-Ray irradiation on powdery chitin produces free radicals on the backbone that initiate polymerization of styrene as in the case of cellulose. Water is essential for initiation, and the grafting percentage is 64% (Shigeno et al., 1982). Photo-induced initiation is also applied to the graft copolymerization of methyl methacrylate onto chitin or oxychitin. A small amount of DMF reduces the induction period and increases both grafting percentage and apparent number of grafted chains. Noncatalytic photo-induced graft copolymerization results in the higher conversion and larger number of grafted chains than photosensitized graft copolymerization with hydrogen peroxide or azobisisobutyronitrile (Takahashi et al., 1989).

Chitin derivatives soluble in organic solvents are important, since they enable modification reactions to proceed in suitable organic solvents under homogeneous conditions. Iodo-chitin is particularly interesting as a precursor because of the solubility in polar solvents and high reactivity (see Chapter 6 in this book), and the graft copolymerization can be carried out in solution or a highly swollen state in organic solvents. On addition of a Lewis acid such as $SnCl_4$ or $TiCl_4$ to iodo-chitin in nitrobenzene, reactive carbenium species are formed, and styrene is efficiently graft-copolymerized by a cationic mechanism in a swollen state (Kurita et al., 1992a):

Table 1 summarizes the results of the graft copolymerization, and the grafting percentage is up to 800%. The grafted chains can be isolated from the copolymers by hydrolytic degradation of the chitin backbone. The polystyrene isolated from a graft copolymer with a grafting percentage of 650% is characterized by M_n, M_w, and M_w/M_n values of 58,000, 87,000, and 1.50. The values of grafting percentage and M_n indicate that each polystyrene chain is attached to every 44 N-acetylglucosamine units on the average. The resulting chitin-*graft*-polystyrenes prepared here are almost soluble in aprotic polar solvents such as dimethyl sulfoxide (DMSO) and N,N-dimethylacetamide containing 5% lithium chloride when the grafting percentage is above 100%. They swell considerably even in common low-boiling organic solvents.

Irradiation of UV light on iodo-chitin in DMSO solution gives rise to the homolysis of the C-I bonds to form carbon free radicals, and styrene is graft-copolymerized by a radical mechanism [Equation (1)].

*Table 1. Cationic and radical graft copolymerization
of styrene onto iodo-chitin.[a]*

Initiation	Styrene (g)	Solvent	Temperature (°C)	Grafting (%)
SnCl₄ (0.10 ml)	3.6	Nitrobenzene (4.0 ml)	10	600
SnCl₄ (0.20 ml)	3.6	Nitrobenzene (4.0 ml)	10	800
SnCl₄ (0.20 ml)	3.6	Nitrobenzene (8.0 ml)	10	115
SnCl₄ (0.20 ml)	3.6	Nitrobenzene (4.0 ml)	5	380
SnCl₄ (0.20 ml)	3.6	Nitrobenzene (4.0 ml)	15	580
UV	0.9	DMSO (2.0 ml)	rt[b]	52
UV	2.7	DMSO (2.0 ml)	rt	63

[a]With SnCl₄: iodo-chitin (ds 0.58), 100 mg; time, 5 h. With UV: iodo-chitin (ds 0.58), 50 mg; time, 1 h.
[b]rt = room temperature.

Although the grafting percentage is not very high (50–60%) as shown in Table 1, no appreciable amount of homopolymer is detected (Kurita et al., 1992a).

Copolymerization of NCAs

To introduce polypeptide side chains into chitin, ring opening graft copolymerization of α-amino acid N-carboxy anhydrides (NCAs) is appropriate. Free amino groups may work as initiators for the graft copolymerization of NCAs, but chitosan or the water-soluble chitin (see Chapter 6 in this book) fails to initiate the reaction in organic solvents on account of the low extents of swelling.

The water-soluble chitin, however, shows high reactivity in aqueous solution, and the graft copolymerization of γ-methyl L-glutamate NCA is accomplished by treating an aqueous solution of the water-soluble chitin with an ethyl acetate solution of the NCA at 0°C (Kurita et al., 1985):

Water-soluble chitin (l ≈ m)

(2)

Even though NCAs are quite susceptible to hydrolysis to give the corresponding amino acids, the grafting efficiency is surprisingly high up to 91% in aqueous media under these conditions (Table 2), and no homopolymerization is observed (Kurita et al., 1988). Graft copolymerization of NCAs onto partially deacetylated chitins is also possible in DMSO, but the grafting efficiency is moderate because of the heterogeneous reaction conditions (Aiba et al., 1985).

The resulting copolymers are a new type of hybrid materials composed of the polysaccharide and polypeptides. Chitin-*graft*-poly(γ-methyl L-glutamate)s are soluble in dichloroacetic acid and hexafluoro-2-propanol and show varying degrees of solubility in common polar solvents, depending on the side-chain length. The graft copolymers having short side chains are almost soluble in DMSO, but the solubility decreases with an increase in the side-chain length. These results are interpreted in terms of two factors: effective destruction of the crystalline structure by incorporation of the short chains and formation of ordered structures of the long side chains including α-helix as suggested by X-ray diffractometry.

The derived membranes exhibit interesting permeation behavior in dialysis, depending also on the side-chain length; urea permeates rapidly when the side chains are short, but the permeation rate decreases with increasing side-chain length.

Ester groups of the side chains are hydrolyzed with sodium hydroxide to give chitin-*graft*-poly(sodium L-glutamate)s that are readily soluble in water [Equation (2)] (Kurita et al., 1988).

The graft copolymerization of NCAs is also interesting in view of introducing polypeptide chains as spacer arms having a terminal free amino group which is utilized for further modifications. Dihydronicotinamide groups have thus been immobilized at the terminals of the side chains of chitin-*graft*-poly(L-alanine) (Kurita et al., 1992b):

Table 2. Graft copolymerization of γ-methyl L-glutamate NCA onto water-soluble chitin.[a]

NCA/GlcN (mol/mol)	Degree of Polymerization[b]		Grafting Efficiency (%)[b]	
	Weight	IR	Weight	IR
2.3	2.1	1.8	91	78
13.6	10.4	10.9	76	80
16.1	13.1	12.3	81	76
26.6	22.0	20.8	83	78

[a]At 0°C for 2 h.
[b]Calculated from weight increase or by IR spectroscopy.

$$\left[\begin{array}{c} \text{HO} \overset{\text{OH}}{\underset{O}{\bigcirc}} O^- \\ \text{NH} + \overset{}{\underset{O}{C}} - \overset{}{\underset{CH_3}{CH}} - \text{NH} + \text{H} \\ \end{array}\right]_n \rightleftharpoons \left[\begin{array}{c} \text{HO} \overset{\text{OH}}{\underset{O}{\bigcirc}} O^- \\ \text{NH} + \overset{}{\underset{O}{C}} - \overset{}{\underset{CH_3}{CH}} - \text{NH} + \overset{}{\underset{O}{C}} \overset{\bigcirc}{\underset{\underset{Bzl}{N}}{}} \\ \end{array}\right]_n \quad (3)$$

The resulting conjugates are models for coenzyme NADH and useful as polymeric asymmetric reducing agents. Actually, incorporation of the spacer arms effectively increases the chemical yield of the reduction of a ketone, indicating the important roles of the polyalanine spacer arms.

GRAFT COPOLYMERIZATION ONTO CHITOSAN

Copolymerization of Vinyl Monomers

Azobisisobutyronitrile initiates graft copolymerization of some vinyl monomers including acrylonitrile, methyl methacrylate, methyl acrylate, and vinyl acetate onto chitosan in aqueous acetic acid solution or in suspension in water. The grafting percentages are, however, generally low. The resulting chitosan-*graft*-poly(vinyl acetate) is converted into chitosan-*graft*-poly(vinyl alcohol) on hydrolysis (Blair et al., 1987). Cerium is another suitable initiator, and polyacrylamide (Kim et al., 1987) and poly(acrylic acid) (Kim et al., 1989) have been introduced. Although chitosan is an effective flocculating agent only in acidic media, the derivatives having poly(acrylic acid) side chains show high-flocculation ability in both acidic and basic regions because of the zwitterionic characteristics (Kim et al., 1989). With Fenton's reagent (Fe^{2+}/H_2O_2) as a redox initiator, methyl methacrylate is graft-copolymerized. The grafting percentage and grafting efficiency are up to 665% and 72% (Lagos and Reyes, 1988).

Graft copolymerization of methyl methacrylate is initiated with UV light. The grafting percentage decreases in the following order: initiator method with ammonium persulfate > photo-induced method without a catalyst > photo-induced method with a photosensitizer. When a mixture of acrylonitrile and styrene is used, almost alternating copolymers ($r_1 \times r_2 = 0.01$) are introduced (Takahashi et al., 1987). γ-Ray irradiation allows graft copolymerization of styrene onto chitosan powders or films. Water promotes grafting as in the case of chitin. The chitosan-*graft*-polystyrene shows higher adsorption of bromine than chitosan, and the films show less swelling and better elongation in water than chitosan films (Shigeno et al., 1981, 1982).

Carbon-carbon double bonds incorporated by acylation of chitosan with maleic anhydride are available for graft copolymerization initiated with ammonium persulfate. The derivative is allowed to react with acrylamide in water to give three-dimensional copolymers:

(4)

The ratio of the double bonds utilized for graft copolymerization is 40–70%. The resulting copolymers are characterized by remarkable swelling in water with a volume increase of 20–150 times (Berkovich et al., 1983).

Copolymerization of Aniline

Polyaniline has been grafted onto chitosan from the viewpoint of preparing conductive polymers. When chitosan in aqueous acetic acid solution is treated with aniline in the presence of ammonium persulfate, polyaniline side chains are introduced at the amino groups (Yang et al., 1989):

(5)

Although chitosan-*graft*-polyaniline gives self-supporting materials such as thick films and fibers, the properties vary depending on the ratio of the amino group to aniline in the grafting reaction. With a ratio of 1:1–1:5, the products are sturdy and flexible, whereas those prepared at a 1:6–1:10 ratio are brittle. Optical microscopic observation indicates that the products prepared at a ratio below 1:5 are homogeneous, but those above 1:6 have crystallites. The copolymers are deep blue and become dark green when treated with hydrochloric acid. The conductivity can be raised from less than 10^{-7} S/cm of the untreated copolymer up to 10^{-2} S/cm on protonic doping.

CONCLUSION

Compared with the graft copolymerization onto cellulose, that onto chitin has been explored only poorly, but it is a rapidly advancing field. Methods for efficient graft copolymerization and analytical procedures for the resulting graft copolymers are currently being developed, and sophisticated molecular design will become possible. The graft copolymers based on chitin will find new applications in various fields such as water treatment, toiletries, medicine, agriculture, food processing, and separation.

REFERENCES

Aiba, S., N. Minoura and Y. Fujiwara. 1985. "Graft Copolymerization of Amino Acids onto Partially Deacetylated Chitin," *Int. J. Biol. Macromol.,* 7:120.

Berkovich, L. A., M. P. Tsyurupa and V. A. Davankov. 1983. "The Synthesis of Crosslinked Copolymers of Maleilated Chitosan and Acrylamide," *J. Polym. Sci., Polym. Chem. Ed.,* 21:1281.

Blair, H. S., J. Guthrie, T. K. Law and P. Turkington. 1987. "Chitosan and Modified Chitosan Membranes. I. Preparation and Characterisation," *J. Appl. Polym. Sci.,* 33:641.

Kim, K. H., K. S. Kim and J. S. Shin. 1987. "Graft Copolymerization of Chitosan and Acrylamide by Using Ceric Ammonium Nitrate Initiator," *Pollimo,* 11:133; *Chem. Abstr.,* 107:9207h, 1987.

Kim, Y. B., B. O. Jung, Y. S. Kang, K. S. Kim, J. I. Kim, and K. H. Kim. 1989. "Flocculation Effect of Chitosan Copolymer Flocculants Grafted with Acrylic Acid," *Pollimo,* 13:126; *Chem. Abstr.,* 111:59884e, 1989.

Kojima, K., M. Yoshikuni and T. Suzuki. 1979. "Tributylborane-Initiated Grafting of Methyl Methacrylate onto Chitin," *J. Appl. Polym. Sci.,* 24:1587.

Kurita, K., M. Kanari and Y. Koyama. 1985. "Studies on Chitin. 11. Graft Copolymerization of γ-Methyl L-Glutamate NCA onto Water-Soluble Chitin," *Polym. Bull.,* 14:511.

Kurita, K., A. Yoshida and Y. Koyama. 1988. "Studies on Chitin. 13. New Polysaccharide/Polypeptide Hybrid Materials Based on Chitin and Poly(γ-methyl L-glutamate)," *Macromolecules,* 21:1579.

Kurita, K., M. Kawata, Y. Koyama and S. Nishimura. 1991. "Graft Copolymerization of Vinyl Monomers onto Chitin with Cerium(IV) Ion," *J. Appl. Polym. Sci.,* 42:2885.

Kurita, K., S. Inoue, K. Yamamura, H. Yoshino, S. Ishii and S. Nishimura. 1992a. "Cationic and Radical Graft Copolymerization of Styrene onto Iodochitin," *Macromolecules,* 25:3791.

Kurita, K., S. Iwawaki, S. Ishii and S. Nishimura. 1992b. "Introduction of

Poly(L-alanine) Side Chains into Chitin as Versatile Spacer Arms Having a Terminal Free Amino Group and Immobilization of NADH Active Sites," *J. Polym. Sci., Part A: Polym. Chem.*, 30:685.

Lagos, A. and J. Reyes. 1988. "Grafting onto Chitosan. I. Graft Copolymerization of Methyl Methacrylate onto Chitosan and Fenton's Reagent," *J. Polym. Sci., Part A: Polym. Chem.*, 26:985.

Lagos, A., M. Yazdani-Pedram, J. Reyes and N. Campos. 1992. "Ceric Ion-Initiated Grafting of Poly(methyl acrylate) onto Chitin," *J. Macromol. Sci., Pure Appl. Chem.*, A29:1007.

Ren, L., Y. Miura, N. Nishi and S. Tokura. 1993. "Modification of Chitin by Ceric Salt-Initiated Graft Polymerisation – Preparation of Poly(methyl methacrylate)-Grafted Derivatives that Swell in Organic Solvents," *Carbohydr. Polym.*, 21:23.

Shigeno, Y., K. Kondo and K. Takemoto. 1981. "Functional Monomers and Polymers. 91. Adsorption of Iodine and Bromine onto Polystyrene-Grafted Chitosan," *Makromol. Chem., Rapid Commun.*, 182:709.

Shigeno, Y., K. Kondo and K. Takemoto. 1982. "Functional Monomers and Polymers. 90. Radiation-Induced Graft Copolymerization of Styrene onto Chitin and Chitosan," *J. Macromol. Sci., Chem.*, A17:571.

Takahashi, A., Y. Sugahara and Y. Horikawa. 1987. "Studies on Graft Copolymerization onto Cellulose Derivatives (Part 28). Graft Copolymerization onto Chitosan by Photo-Induced and Initiator Methods," *Sen'i Gakkaishi*, 43:362; *Chem. Abstr.*, 107:97249b, 1987.

Takahashi, A., Y. Sugahara and Y. Hirano. 1989. "Studies on Graft Copolymerization onto Cellulose Derivatives. XXIX. Photoinduced Graft Copolymerization of Methyl Methacrylate onto Chitin and Oxychitin," *J. Polym. Sci., Part A: Polym. Chem.*, 27:3817.

Yang, S., S. A. Tirmizi, A. Burns, A. A. Barney and W. M. Risen. 1989. "Chitaline Materials: Soluble Chitosan-Polyaniline Copolymers and Their Conductive Doped Forms," *Synth. Met.*, 32:191.

PART VI

WASTEWATER TREATMENT

CHAPTER 18

Applications of Chitosan for the Elimination of Organochlorine Xenobiotics from Wastewater

J. P. THOME AND CH. JEUNIAUX

University of Liège
Laboratory of Animal Ecology and Ecotoxicology
22 Quai Van Beneden
B-4020 Liège, Belgium

M. WELTROWSKI

Textile Technology Centre
Rue Boullé 3000
St-Hyacinthe, Québec, Canada J2S 1H9

INTRODUCTION

In industrial areas, large volumes of process water, containing organic and inorganic contaminants, are discharged in aerated lagoons or aerated sludge systems. In these waste treatment facilities persistent organic micropollutants like organochlorine hydrocarbons remain in fixed sediments or on suspended matter for a long time because of their resistance to physical, chemical, and biological degradation. Standard methods for water purification used in water-softening plants remain largely ineffective to remove these persistent xenobiotics. However, the adsorption process could be a good method for the elimination of these organic pollutants from the aquatic environment. Several studies surveyed the binding of toxic organic compounds, such as pesticides, by various types of synthetic resins, biopolymers such as cellulosic derivatives, and activated charcoal (Koshijima et al.,

309

1973; Tucker and Saeger, 1975; Davar and Wightman, 1981). Chitin and especially its deacetylated derivative, chitosan, have found numerous applications in industrial processes and especially in wastewater treatment and in potable water purification (Muzzarelli and Tanfani, 1981).

Chitosan has been used as an effective flocculent and coagulating agent for organic matter (Bough et al., 1976), as an effective agent for coagulation of suspended solids, and for recovering proteins from various food-processing wastes (Bough, 1975, 1976; Bough and Landes, 1978; Inoue et al., 1988). Moreover, this polymer reveals interesting binding properties for dye adsorption (McKay et al., 1982; Knorr, 1983; Muzzarelli, 1973; Weltrowski et al., 1996). Chitinous materials and especially chitosan and its derivatives have provided a new class of functional polymeric materials as gels for enzyme immobilization, supports of liquid chromatography, and selectively permeable membranes (Seo et al., 1991).

The use of microorganisms for biodegradation of toxic chemicals has pointed out novel applications for chitin and chitosan in the field of wastewater epuration. Chitin was used as a solid matrix for attachment of bacterial strains involved in detoxification and biodegradation processes of various organic micropollutants, e.g., chlorophenols (Portier, 1986, 1989; Portier and Fujisaki, 1986).

On the basis of the unique structure and properties of chitin-deacetylated derivatives, high-performance adsorbents based on chitosan seemed to be promising (Kurita et al., 1988). Several studies report that the chemical modification of chitosan was performed to improve the sorption capacity of this polymer (Seo and Iijima, 1991). These authors investigated in detail the interactions of chitosan gels containing hydrophobic groups with small hydrophilic and hydrophobic molecules. They emphasized the significant contribution of hydrophobic interaction in the mechanism of binding.

It has been shown that the metal-binding capacities of chitosan were improved by a chemical modification of the amino group (Muzzarelli and Tanfani, 1982; Muzzarelli et al., 1985; Keisuke et al., 1988) and/or the cross-linking of chitosan by means of glutaraldehyde (Masri, 1980; Masri and Randall, 1978; Yoshiyaki and Akihiko, 1986; Keisuke et al., 1986).

USE OF CHITOSAN TO RECOVER
ORGANOCHLORINATED COMPOUNDS

Metal-binding capacities of unmodified chitosan and chitosan derivatives are well documented. However little research has been done on the interactions of chitin and chitosan with organochlorinated persistent contaminants, such as pesticides [e.g., pentachlorophenol (PCP)] and PCBs (polychlorinated biphenyls). Richards and Cutkomp (1945) followed by Lord (1948) showed that chitin exhibits sorption properties for an organochlorinated pesticide, DDT, and related compounds. Davar and Wightman (1981) reported that among organochlorinated pesticides, only acidic pesticides, such as 2,4-D, 2,4,5-T, Dicamba, and MCPA, showed significant uptake on the chitin-deacetylated derivative, chitosan. These authors suggested that the sorption isotherms of these acidic pesticides on chitosan revealed a "side-by-side" association between the adsorbed molecules, helping to hold them to the surface. Giles et al. (1974) designed this type of "multilayer adsorption" as "cooperative adsorption" with the solute molecules tending to be adsorbed, packed in rows or clusters. Other mechanisms are suggested for the sorption of organochlorine compounds onto activated charcoal. On this substrate, the sorption isotherm of 2,4-D showed that as more sites in the charcoal are filled, it becomes increasingly difficult for a solute molecule to be adsorbed on a vacant site. According to Davar and Wightman (1981), this is indicative of "monolayer adsorption" of molecules. Therefore, these authors concluded that, in contrast to the sorption process involved with chitosan, adsorption on the charcoal is a "pore-filling" process, in which the available volume of the pore is the controlling factor. Thus, the active sites in chitosan are believed to be the amino groups, and acidic pesticides are fixed by means of ionic sorption process rather than physical adsorption, which involves a surface phenomenon. The ionic sorption takes place through electrostatic interaction of the carboxyl anion of pesticides and the protonated amine site of chitosan.

A CLASS OF PRIORITY ORGANOCHLORINE POLLUTANTS: THE PCBs

PCBs are synthetic organic molecules that have been widely used in various industrial sectors (e.g., plastics, electricity, lubricants, and hydraulic systems), notably to develop, combined with other organic substances, insulating fluids such as Askarel® and Pyralene® used in electric transformers and capacitors. It is theoretically possible to produce 209 different congeners, differing by the number and position(s) of chlorine atoms on the biphenyl ring. The PCB molecules are divided into ten isomeric groups (monochlorine to decachlorine). PCBs have always been used in complex mixtures of congeners characterized by their different chlorine contents (Hützinger et al., 1974). PCB mixtures are sold under trade names, such as Aroclor® (Monsanto, USA),

Clophen® (Bayer, Germany), and Phenoclor® (France). The Aroclor mixtures, especially Aroclor 1254 and Aroclor 1260 (respective chlorine contents: 54% and 60% by weight), have been abundantly used in Europe and North America.

According to their high lipophilicity, their low degradability by chemical, microbial, and metabolic transformations, PCBs accumulate in living organisms and up the food chain by the processes of biomagnification and bioconcentration in terrestrial and aquatic ecosystems. As a consequence, the PCBs, and especially the Aroclor-derived PCBs, are responsible for an ubiquitous contamination of the global ecosystems. In spite of legal regulations and controls on PCB use, these xenobiotics continue to enter the aquatic environment by means of atmospheric transport and from PCB contaminated wastewater or effluent originating mainly from discharges (Waid, 1986; Duursma and Marchand, 1974).

PCBs accumulate in marine and freshwater crustacean planktonic organisms because of their primarily hydrophobic chemical properties. On the other hand, PCBs can easily adsorb on organic particulate matter and on living and dead planktonic organisms (Duursma and Marchand, 1974; Clayton et al., 1977; Hiraizumi et al., 1979). According to the affinity of chitin for PCBs, especially in crustacean planktonic cuticle (Hiraizumi et al., 1979; Thomé et al., 1991), one can suggest that PCB adsorption would be a main process in contamination of crustaceans belonging to the marine and freshwater zooplankton. As a consequence, chitin and its derivatives must likely constitute a promising PCB-adsorbing agent useful for the scavenging (i.e., purification) of contaminated stream water.

The use of PCBs in complex mixtures generates major difficulties in the development of a successful elimination process of these xenobiotics from contaminated natural and waste waters according to the different physicochemical characteristics of the PCB isomers and congeners. Moreover, studies performed with individual PCB congeners indicate that PCB toxicity depends on congener structure. This notably applies to toxicity involving, e.g., induction of hepatic cytochrome P-450 monooxygenase (McFarland and Clarke, 1989; Safe, 1990; Sparling and Safe, 1980). In this respect, the most potent PCBs are the few coplanar ones (i.e., IUPAC No. 77, 126, and 169), which sterically resemble 2,3,7,8-tetrachloro dibenzo-p-dioxin (2,3,7,8-TCDD), and a handful of partially coplanar forms (IUPAC No. 105, 118, 128, 138, 156, and 170). Given the likelihood of toxicity to animals and the prevalence of individual PCB congeners in the environment, it is now generally believed that only 15–20 congeners are of real concern (McFarland and Clarke, 1989; Duinker et al., 1988; Sariaslani, 1991). Most of the above-mentioned highly toxic PCB congeners are present in Aroclor mixture (Aroclor 1254 and Aroclor 1260).

Standard methods for water purification used in water-softening plants remain largely ineffective for eliminating highly persistent organic xenobiotics. Until the 1980s, expensive treatments, leading only to an incomplete elimination, such as adsorption onto activated charcoal or synthetic resins, have been used (Tucker and Saeger, 1975). The high affinity of organochlorine hydrocarbons to chitin and chitosan, observed in vitro in the beginning of the 1980s, has instigated new research to develop effective PCB-contaminated stream water treatment using these natural polymers. Because of the ubiquitous presence of Aroclor-derived PCBs in the environment, the development of such an original methodology for PCB removal from contaminated wastewater must not solely take into account the most toxic PCB congeners but also the major part of PCB-pure components constituting the Aroclor commercial mixture.

PRELIMINARY STUDIES OF PCB SORPTION ON CHITIN AND CHITOSAN

Thomé and Van Daele started these studies by testing the chitin and chitosan efficiency for removing PCBs from contaminated water (Thomé and Van Daele, 1986; Van Daele and Thomé, 1986). Chitin was extracted from crude shrimp shells (*Crangon crangon*) according to the procedure described by Jeuniaux (1963), and chitosan was prepared as described by Asano (1978). The PCB adsorption efficiency on the adsorbing agent was tested by filtration of distilled water spiked with environmental concentrations (0.5 μg/L) of a PCB mixture (Aroclor 1260) through cartridges filled with the adsorbents. The PCB removal efficiency from water was 83.3 ± 7.5%, 66.1 ± 33%, and 66.6 ± 1.2% for chitosan, chitin, and activated carbon, respectively. When compared with activated carbon, chitin had similar PCB adsorption properties but appeared less effective than chitosan. Thomé and Van Daele (unpublished data) showed that the PCB adsorption isotherm on chitosan was similar to that obtained by Davar and Wightman (1981) with acidic pesticides. The initial curvature of this isotherm suggests the so-called "co-operative adsorption" as described by Giles et al. (1974) and could be attributed to multilayer adsorption. Nevertheless, these data were insufficient to consider that the mechanism of PCB sorption on chitosan was elucidated. Moreover, Van Daele and Thomé (1986) showed that PCB removal from contaminated water at environmental concentrations (μg/L level) was not complete but quite sufficient to decrease the PCB contamination to beneath the "toxic threshold level" and to effectively protect fish (*Barbus barbus*) against serious metabolic diseases that affect growth and reproduction. These results suggest that the scavenging of PCB-contaminated water using chitosan could find favorable application in environmental preservation. However, although the partition coefficient of PCBs between chitosan and water is very important, unmodified chitosan leads only to incomplete removal of PCB from natural or wastewater. According to Thomé and Van Daele (1986) and Van Daele and Thomé (1986), the concentration factor of PCBs for chitosan reached up to 2.5×10^4.

PCB SORPTION ABILITY OF CHEMICALLY MODIFIED CHITOSAN

To improve the PCB sorption properties of chitosan, Thomé et al. (1992) chemically modified the amino group of chitosan. Various chitosan derivatives were synthesized by means of a cross-linking procedure with benzoquinone and glutaraldehyde (Figure 1): glutaraldehyde cross-linked chitosan (CHT-Glu), glutaraldehyde cross-linked NaBH$_3$CN reduced chitosan (CHT-Glu-Red), and benzoquinone cross-linked chitosan (CHT-BQ).

These authors tested the efficiency of the chitosan derivatives by means of batch and flow-through chitosan cartridge systems using a PCB mixture (Aroclor 1260) whose ubiquitous presence in the aquatic

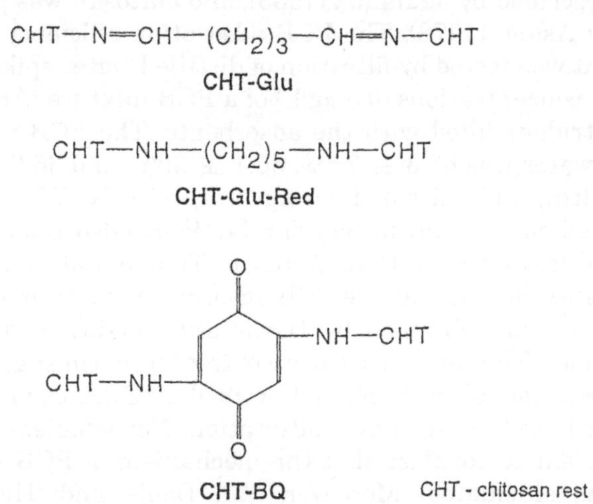

$$CHT-N=CH-(CH_2)_3-CH=N-CHT$$
CHT-Glu

$$CHT-NH-(CH_2)_5-NH-CHT$$
CHT-Glu-Red

CHT-BQ   CHT - chitosan rest

Figure 1. Chitosan derivatives synthesized by cross-linking with benzoquinone (CHT-BQ), glutaraldehyde (CHT-Glu) and secondarily reduced by NaBH$_3$CN (CHT-Glu-Red) (from Thomé et al., 1992.)

Table 1. *PCB (Aroclor 1260) adsorption efficiency on various chitosan derivatives in batch system (24 h of PCB/chitosan contact) (Thomé et al., 1992).*

Initial PCB Conc. in Water	% of PCBs (Aroclor 1260) Removed from Contaminated Water			
	CHT[b]	CHT-Glu-Red[c]	CHT-Glu[d]	CHT-BQ[e]
1 μg/L	93.0	94.4	70.7	48.3
10 μg/L[a]	96.9	96.5	87.6	89.1
100 μg/L[a]	96.2	94.0	59.3	50.8
1,000 μg/L[a]	39.9	82.3	73.3	58.8

[a]These concentrations are higher than the maximum hydrosolubility of Aroclor 1260 (5.4 μg/L at 20°C).
[b]Unmodified chitosan.
[c]Glutaraldehyde cross-linked NaBH$_3$CN reduced chitosan.
[d]Glutaraldehyde cross-linked chitosan.
[e]Benzoquinone cross-linked chitosan.

environment was demonstrated. The Aroclor 1260 is a PCB mixture containing highly chlorinated PCB individual components (mainly hexa- and hepta-CB).

Efficiency of PCB Binding Ability of Chitosan Derivatives in Batch System

The efficiency of PCB binding abilities of the chemically modified chitosan was tested in a batch system by Thomé et al. (1992) (Table 1). The unmodified chitosan (CHT) and the glutaraldehyde cross-linked derivative secondarily reduced by NaBH$_3$CN (CHT-Glu-Red) exhibited high rates of adsorption that generally exceeded 90%. However, the two other chitosan derivatives cross-linked either with glutaraldehyde (CHT-Glu) or with p-benzoquinone (CHT-BQ) appeared dramatically less effective. Because of the water solubility of the PCB mixture used (Aroclor 1260, 5.4 μg/L at 20°C), Thomé et al. (1992) presumed that, for the less efficient chitosan derivatives, i.e., chitosan cross-linked with glutaraldehyde or with benzoquinone (CHT-BQ and CHT-Glu), a competition occurred between the glassware walls and the chitosan for PCB sorption because of the poor affinity of these two chitosan derivatives for PCBs and the resulting relatively long residual time of PCBs in water.

Efficiency of PCB Binding Ability of Chitosan Derivatives in Flow-Through System

In a flow-through system, unmodified chitosan as well as chitosan cross-linked with glutaraldehyde and secondarily reduced by NaBH$_3$CN appeared to have very high capacities for PCB binding (Thomé et al., 1992). The chitosan cross-linked with benzoquinone and glutaraldehyde remained largely less effective (up to 50% of PCBs remaining in effluent after filtration through the chitosan derivatives) (Table 2).

The relative importance of individual PCB components in residual water is quite similar whatever the nature of the two most efficient chitosan derivatives (CHT and CHT-Glu-Red) as illustrated in Figure 2. The PCB101, PCB194, and PCB195 (i.e., penta- and hepta-CBs) were totally removed from filtered water when the treatment was done with these two chitosan derivatives.

The difference between the batch system and the column system may be explained by the different contact time between the PCB and the polymer molecules. In the batch system the residual time of PCBs in water is sufficiently long for balancing the differences between the adsorption efficiency of the chitosan derivatives. In the column system, the time is considerably shorter. As a consequence, only the most effective compounds are capable of registering high adsorption rates. Based on these results and on earlier observations of Davar and Wightman (1981), Thomé et al. (1992) hypothesized that the presence of the primary and secondary amino groups is essential for the PCB binding properties of the chitosan derivatives adsorption. As the NH_2 groups disappear in the CHT-Glu, the adsorption properties decrease significantly. The cross-linked derivative CHT-BQ has a secondary amino group, but the steric hindrance caused by the high cross-linkage rate would make the derivative less effective. The results of filtration in the column system show that the derivative CHT-Glu-Red is more effective than the noncross-linked chitosan (CHT). This result may be explained by the increase in the hydrophobic character of the CHT-Glu-Red derivative compared with the unmodified chitosan. Through the glutaraldehyde cross-linking and the secondary $NaBH_3CN$ reduction of chitosan, the primary amino group was changed to a secondary amino group, and a hydrophobic functional group in the form of the aliphatic chain of five

Table 2. PCB (Aroclor 1260) adsorption efficiency on various chitosan derivatives in flow-through system (Thome et al., 1992).

Initial PCB Conc. in Water	% of PCBs (Aroclor 1260) Removed from Contaminated Water			
	CHT[a]	CHT-Glu-Red[b]	CHT-Glu[c]	CHT-BQ[d]
5 μg/L	96.8	99.4	51.4	65.2

[a]Unmodified chitosan.
[b]Glutaraldehyde cross-linked $NaBH_3CN$ reduced chitosan.
[c]Glutaraldehyde cross-linked chitosan.
[d]Benzoquinone cross-linked chitosan.

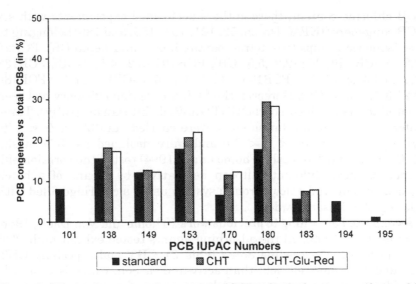

Figure 2. Relative importance (in percent) of PCB individual congeners (designed by their IUPAC number, PCB101 = tetra-CB; PCB138, PCB149, PCB153 = hexa-CB; PCB170, PCB180, PCB183 = hepta-CB; PCB194, PCB195 = octa-CB) remaining in water after filtration through chitosan derivatives (CHT: unmodified chitosan; CHT-Glu-Red: NaBH₃CN reduced glutaraldehyde cross-linked chitosan) (modified from Thomé et al., 1992).

CH$_2$ units was introduced. Moreover, concerning this derivative (CHT-Glu-Red), even if the cross-linked ratio is fairly high, the steric hindrance does not impede adsorption, because the aliphatic chain $-(CH_2)_5-$ ensures a sufficient distance between the polysaccharide chains (Thomé et al., 1992).

PCB SORPTION ON CROSS-LINKED CHITOSAN AS A FUNCTION OF THE MESH SIZE AND THE PCB CHLORINE CONTENT

It has been shown that the cross-linkage of chitosan by hydrophobic functional groups such as aldehyde increased the PCB sorption efficiency on chitosan (Thomé et al., 1992). To develop filter cartridges useful for the scavenging (i.e., purification) of contaminated stream water, Thomé et al. (1994) investigated the PCB adsorption efficiency of two chitosan derivatives synthesized by means of a cross-linking procedure with glutaric aldehyde (CHT-Glu-Red) and terephthaldehyde (CHT-Ter-Red) secondarily reduced by NaBH₃CN, on activated carbon used as a reference material and on pure chitosan (CHT). To evaluate the sorbing efficiency of the chitosan derivatives as a function of the

PCB chlorine content, these authors performed experiments with six PCB congeners (IUPAC No. 28, 52, 101, 138, 153, and 180) belonging to five isomeric groups (tri-, tetra-, penta-, hexa-, and hepta-CB): PCB28 (2,2′,4′-CB), PCB52 (2,2′,5,5′-CB), PCB101 (2,2′,4,5,5′-CB), PCB138 (2,2′,3,4,4′,5′-CB), PCB153 (2,2′,4,4′,5,5′-CB), and PCB180 (2,2′,3,4,4′,5,5′-CB). Moreover, the PCB adsorption efficiency has been related to the mesh size of the CHT-Glu-Red chitosan derivative. Three different granulations (mesh sizes) were carried out ($325 < G < 170$; $170 < G < 100$; $100 < G < 70$) and were designed as "325 mesh," "170 mesh," and "70 mesh." Thomé et al. (1994) tested the sorption efficiency of these different chitosan derivatives by means of a batch (closed system) and a flow-through system using a cartridge filled with 100 mg of chitosan powder.

As expected according to previous work (Thomé and Van Daele, 1986; Thomé et al., 1992), all the adsorbing agents tested exhibit high PCB adsorption rates exceeding 85% (Table 3). CHT and especially CHT-Glu-Red were more efficient than activated carbon, which is generally considered as the most powerful adsorbent for organochlorinated micropollutants. Moreover, the PCB adsorption efficiency of CHT-Glu-Red appeared mesh size independent. The activated carbon, CHT, and CHT-Glu-Red showed the same adsorption efficiency for the various PCB congeners whatever their chlorine content. In contrast, the PCB adsorption efficiency of CHT-Ter-Red increased with the chlorine content of PCB congeners. Therefore, penta-, hexa-, and hepta-CB (i.e., PCB101, PCB138, PCB153, and PCB180) showed higher affinity for this specific cross-linked chitosan than the lower chlorinated ones.

Table 3. PCB adsorption rates (%) on various adsorbing agents in batch system (Thomé et al. 1994).

Adsorbent	PCB28	PCB52	PCB101	PCB153	PCB138	PCB180
Activated carbon	86.1%	85.6%	85.8%	85.1%	85.7%	86.3%
CHT[a]	91.8%	90.3%	91.2%	88.6%	90.5%	90.5%
CHT-Glu-Red[b] "70 mesh"	100%	100%	100%	100%	100%	100%
CHT-Glu-Red[c] "325 mesh"	100%	100%	100%	96.2%	100%	97.7%
CHT-Ter-Red[d]	73.1%	71.5%	81.8%	82.5%	86.4%	88.0%

[a]Unmodified chitosan.
[b]Glutaraldehyde cross-linked NaBH₃CN reduced chitosan (mesh size $100 < G < 70$).
[c]Glutaraldehyde cross-linked NaBH₃CN reduced chitosan (mesh size $325 < G < 170$).
[d]Terephthaldehyde cross-linked NaBH₃CN reduced chitosan.
Contact time: 24th; initial concentration 1 μg/L for each PCB congener.

Table 4. *Recovery of the various PCB congeners bonded on chitosan (CHT) and on cross-linked and secondarily reduced chitosan (CHT-Glu-Red) according to the extraction procedure (in % of the total amount of PCB removed from contaminated water [initial conc. 1 µg/L of each PCB congener) (Thomé et al., 1994).*

PCB Congeners	Liquid/Liquid Extraction		NaOH Saponification	
	CHT	CHT-Glu-Red	CHT	CHT-Glu-Red
PCB28	8.9%	100%	80.7%	0%
PCB52	2.3%	100%	85.2%	0%
PCB101	6.6%	100%	82.9%	0%
PCB138	15.2%	96.3%	75.4%	3.7%
PCB153	12.5%	96.6%	76.0%	3.2%
PCB180	16.3%	94.5%	73.8%	3.1%

CHT-Ter-Red, i.e., chitosan derivative cross-linked with aromatic carbon chain, appeared less efficient for the PCB fixation than CHT-Glu-Red cross-linked with aliphatic carbon chain.

The PCB adsorption mechanism on chitosan appeared quite different according to the nature of the chitosan derivative. Thomé et al. (1994) demonstrated that the PCBs adsorbed on pure chitosan (CHT) and on cross-linked chitosan (CHT-Glu-Red) have to be recovered by using a very different extraction procedure as shown in Table 4. Indeed, an organic solvent liquid extraction procedure based on the hydrophobic chemical properties of PCBs, allowed for the recovery of high amounts of adsorbed PCBs on CHT-Glu-Red chitosan derivative. However, this method remained largely ineffective to recover the PCBs theoretically fixed on unmodified chitosan (CHT). The use of a more vigorous PCB extraction involving an NaOH-saponification procedure led to the recovery of the PCBs fixed onto chitosan by strong ionic interaction.

These results showed that, with few differences according to the PCB chlorine content, PCB sorption on two chitosan derivatives (CHT and CHT-Glu-Red) is the result of a combination of two kinds of binding processes: strong ionic bonds especially on unmodified chitosan (CHT) and weak nonionic hydrophobic interactions predominating in PCB adsorption on cross-linked chitosan (CHT-Glu-Red).

Thomé et al. (1994) showed the breakthrough curves for the filtration system of cartridges filled with 100 mg of chitosan derivatives (i.e., CHT, CHT-Glu-Red "70 mesh," and CHT-Glu-Red "170 mesh") and activated carbon. The PCB sorption efficiency of the different sorbing agents appeared quite different according to the results obtained for three congeners that are representative of their respective isomeric class, i.e., PCB52 (tetra-CB), PCB101 (penta-CB), and PCB180 (hepta-CB) (Figure 3).

The results presented in Figure 3 clearly show the high affinity of all adsorbing agents tested for PCBs. However, the reduced cross-linked chitosan (CHT-Glu-Red) exhibited the highest PCB binding efficiency. The CHT-Glu-Red "170 mesh" appeared as the only adsorbent that kept its maximum PCB adsorption efficiency for all the PCB congeners whatever their chlorine content. The other adsorbing agents (CHT, CHT-Glu-Red "70 mesh," and the activated carbon) became rapidly ineffective (only about 60–80% of PCB bonded on adsorbent) for hepta-CB (PCB180). Moreover, the unmodified chitosan is less efficient than the activated carbon.

In short, in flow-through systems, the PCB adsorption efficiency of CHT-Glu-Red remained independent of the chlorine content of PCB congeners and was related to the mesh size (the lowest particle size increased the PCB sorption efficiency) (Thomé et al., 1994). These results are in good agreement with Hiraizumi et al. (1979) who suggested that the capacity of PCB adsorption onto suspended matter in the marine environment was mainly dependent on particle size.

EVALUATION OF REDUCED CROSS-LINKED CHITOSAN FOR EPURATION OF LARGE VOLUMES OF PCB-CONTAMINATED STREAM WATER

PCB Sorption Capacity of Reduced Cross-Linked Chitosan (CHT-Glu-Red) Powder Filter

In recent work, Thomé et al. (1996) performed the final evaluation of CHT-Glu-Red chitosan derivative's efficiency for PCB removal from large volumes (up to 40 L) of highly contaminated water (10–50 μg/L of each PCB congener). Experiments performed by these authors involved, on the one hand, cartridges filled with CHT-Glu-Red derivative powder in a flow-through system and, on the other hand, a nonwoven filter (K8WOT filter), based on cross-linked chitosan, developed and tested for PCP sorption by Weltrowski et al. (1994).

According to the results obtained by Thomé et al. (1994) and Weltroswki et al. (1994), the optimal particle size of the CHT-Glu-Red for high PCP and PCB adsorption efficiency is the granulation range $170 < G < 100$ mesh. This granulation was chosen by Thomé et al. (1996) for testing the PCB101 (penta-CB) adsorption efficiency for large volumes of contaminated water (40 L) by means of a flow-through column system.

The breakthrough curve for the PCB101 filtration is presented in Figure 4. The initial concentration was 50 μg/L, i.e., the upper limit of

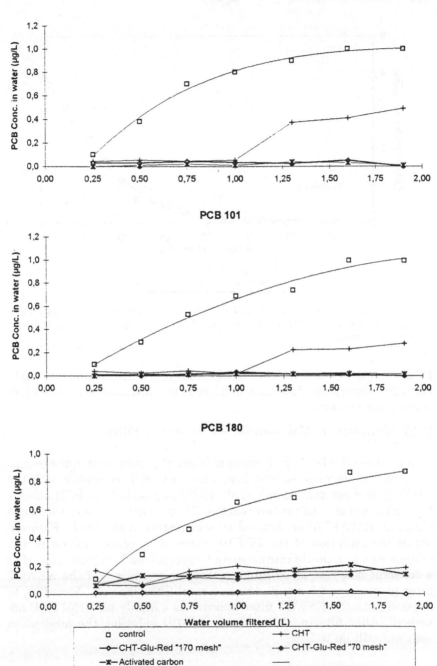

Figure 3. PCB concentrations remaining in residual contaminated water after filtration through various adsorbents as a function of the volume filtered (modified from Thomé et al., 1994).

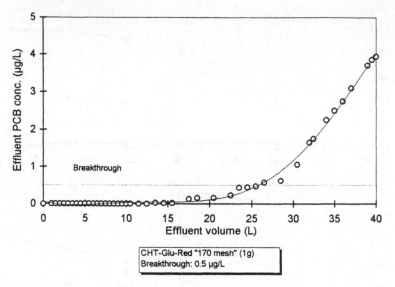

Figure 4. Breakthrough curves obtained for filtration of PCB101 contaminated water (50 μg/L) through a 1 g CHT-Glu-Red column (170 < G < 100 mesh) (modified from Thomé et al., 1996).

PCB101 hydrosolubility. The breakthrough concentrations (0.5 μg/L, i.e., a 99% adsorption rate) were achieved after filtration of 25 L of PCB-contaminated water.

PCB Adsorption on Chitosan-Based Nonwoven Filter

To increase the facility of manipulation, the treatment space reduction, and the decrease of lead loss, which occured inevitably with the CHT-Glu-Red powder, Thomé et al. (1996) also tested the PCB adsorption efficiency of a nonwoven reactive filter based on the cross-linked chitosan (K8WOT filter) devised by Weltrowski et al. (1994). Figure 5 shows the evolution of the PCB101 (penta-CB) concentration in the effluent as a function of contaminated water (50 μg/L) volume passed at a constant flow rate of 20 mL/min through ten layers of the K8WOT filter placed in a 3.5-cm ∅ stainless steel filtration cell. According to these results, the K8WOT filter appears as a highly powerful PCB adsorbent. After filtering 13 L of the PCB101 solution, the adsorption rate was still up to 95%.

PENTACHLOROPHENOL (PCP) ADSORPTION ON
CHITOSAN DERIVATIVES

Pentachlorophenol (PCP) is a pesticide widely used in wood treatment against fungi. With time the PCP is slowly liberated from the wood and contaminates nearby water systems. This problem is observed especially in areas where treated wood is stocked. As with PCBs, PCP is a lipophilic persistent organochlorinated xenobiotic resistant to chemical and biological degradation, which accumulates in living organisms and up food chains by the bioaccumulation process.

Adsorption on activated carbon is considered as the most effective method to eliminate pesticides from water. This was shown for different pesticides (Hall et al., 1992; Urano et al., 1991; Toussant and Parent, 1986) and for PCP (Toussant and Parent, 1986). The adsorption system on activated carbon is highly efficient. The disadvantage of this method is the expensive and time-consuming regenerating process. Only the thermal desorption at high temperature (800°C) is applicable. The spent carbon has to first be removed from the system and sent to a special treatment facility. Therefore, this process can be justified only for a large-scale installation.

As it was previously presented, chitosan and its derivatives are powerful adsorbents of various organic xenobiotics, such as organochlorinated pesticides and PCBs. Therefore, the chitosan derivatives useful for PCB adsorption, described by Thomé et al. (1992, 1994, 1996), were tested for PCP adsorption by Weltrowski et al. (1993, 1994a, 1994b). First, the most powerful chitosan derivative was identified by a filtra-

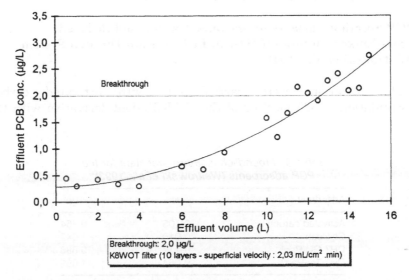

Figure 5. Breakthrough curves obtained for the filtration of PCB101 contaminated water (50 μg/L) through ten layers of the K8WOT filter (modified from Thomé et al., 1996).

tion experiment in a batch system. Second, after the optimization of the granulation of the chitosan derivative, the small and large volume filtrations were conducted. Finally, a chitosan-based nonwoven filter (K8WOT) was developed (Weltrowski et al., 1994a) and tested for PCP adsorption.

Identification of the Most Effective Chitosan-Based PCP Adsorbent

The identification of the most useful adsorbent for PCP sorption was determined by the adsorption isotherms according to the Freundlich model (Weltrowski et al., 1993). Four potential sorbents were tested: activated carbon, unmodified chitosan (CHT), gluteraldehyde cross-linked and NaBH$_3$CN reduced chitosan (CHT-Glu-Red); and terephthaldehyde cross-linked and NaBH$_3$CN reduced chitosan (CHT-Ter-Red). The last two adsorbents were obtained as described by Thomé et al. (1992, 1994). The Freundlich isotherm constant K and $1/n$ are presented in Table 5.

The K constant shows the adsorption capacity and the $1/n$ constant the sorption intensity. For activated carbon the sorption capacity (K) is over 6 times higher than that of the second-best adsorbent CHT-Glu-Red. These results point out that in a static batch system activated carbon is the best PCP adsorbent. However, the CHT-Glu-Red chitosan derivative presents an interesting sorption capacity at over 10 times higher than that of the unmodified chitosan.

PCP Sorption Capacity of Activated Carbon and Reduced Cross-Linked Chitosan (CHT-Glu-Red) in Flow-Through System (Weltrowski et al., 1994)

The filtrations were carried out with activated carbon (granulation < 325 mesh) and the CHT-Glu-Red derivative with the

Table 5. Freundlich isotherm constant for the PCP adsorbents (Weltrowski et al., 1993).

Adsorbent	K	$1/n$
Activated carbon	0.705	0.254
Chitosan	0.010	1.042
CHT-Glu-Red[a]	0.117	0.686
CHT-Ter-Red[b]	0.054	1.067

[a]Glutaraldehyde cross-linked NaBH$_3$CN reduced chitosan.
[b]Terephthaldehyde cross-linked NaBH$_3$CN reduced chitosan.

Table 6. Breakthrough volumes for 5,000 μg/L PCP
solutions filtered through two 1-g columns.

Adsorbent	Maximum Filtration Flow Rate (mL/min)	Breakthrough Volumes[a] (mL)	
		First Column	Second Column
Activated carbon	0.6	3,300	8,000
CHT-Glu-Red[b]	2.8	1,600	7,000

[a]Breakthrough concentration 20 μg/L (99.6% adsorption efficiency).
[b]Glutaraldehyde cross-linked NaBH₃CN reduced chitosan.

same granulation. The 5,000 μg/L PCP solutions were filtered through a system of two minicolumns (0.286 cm²), each one loaded with 1 g of adsorbent.

The breakthrough concentration was established at 20 μg/L. The obtained breakthrough volumes are presented in Table 6. In this dynamic system the PCP adsorption efficiency of the activated carbon and the CHT-Glu-Red chitosan derivative are comparable. The breakthrough volumes for the second column are very close. Moreover, the result for the chitosan derivative was obtained at over a 5 times higher flow rate. The difference in PCP adsorption efficiency in static and dynamic systems could be explained by the nature of the adsorption attraction forces. In the case of the activated carbon, pure physical adsorption occurred. To achieve the maximum adsorption, much more time is required. The PCP sorption on the chitosan derivative probably is achieved by a weak chemical interaction, for example, dipole-dipole attraction. This process is faster and more adapted to the dynamic configuration of the flow-through system. This hypothesis is confirmed by the specific surface data. The specific surface is 835 m²/g for the activated carbon and less than 2 m²/g for the chitosan derivative (Weltrowski, unpublished data). These values show clearly that PCP cannot be adsorbed on the chitosan derivative by physical adsorption. The PCP desorption results confirm the hypothesis of the existence of two different adsorption mechanisms. The desorption rates achieved with 0.01 M NaOH for activated carbon and the CHT-Glu-Red derivative were, respectively, 25% and 77%. The PCP blocked in the activated carbon pores by physical adsorption could not be desorbed by the weak base, in spite of the very high affinity of the PCP to the basic solution. In the case of the chitosan derivative, the weak chemical attraction forces, such as dipole-dipole interactions, could be easily destroyed by the ionic interactions existing between a strong base (NaOH) and a weak acid (PCP).

The flow-through experiments were finished by the filtration of a 5,000 µg/L PCP solution through the 3.28 cm² column loaded with 5 g of CHT-Glu-Red (170 < G < 100 mesh granulation). This new optimized granulation achieved a high-filtration flow rate (30 ml/min) and, moreover, a very high breakthrough volume of 127 L (20 µg/L breakthrough concentration). The activated carbon (G < 325 mesh) turned rapidly unusable due to very high head losses and low flow rates. The desorption rate for the CHT-Glu-Red derivative column, achieved with a 0.01 M NaOH solution, was 94%.

Although the adsorption rates in the dynamic system are, for both sorbents, very similar, the very high desorption rates achieved for the CHT-Glu-Red derivative make it more useful for a PCP adsorbent than the activated carbon that needs an expensive thermal desorption.

PCP Adsorption on the Chitosan-Based Nonwoven Filter

The chitosan-based K8WOT filter developed by Weltrowski et al. (1994a) had its preliminary test for PCP adsorption. A PCP solution (5,000 µg/L) was passed through ten layers of K8WOT filter placed in a 9.62-cm² filtration cell. The flow rates were 19.4 mL/min, and the breakthrough concentration was fixed at 20 µg/L. The breakthrough volume obtained for this filtration system was over 16 L (Weltrowski et al., 1994b). This result shows that the K8WOT filter is a powerful PCP adsorbent which can be easily regenerated with a 0.01 M NaOH solution (Weltrowski, unpublished data). The K8WOT filter also has other advantages, such as facility of manipulation and regeneration, treatment space reduction, and low head losses.

CONCLUSIONS

In this synthesis, it was clearly demonstrated that chitosan (CHT) and especially glutaraldehyde cross-linked and secondarily $NaBH_3CN$ reduced chitosan (CHT-Glu-Red) constitute powerful adsorbing agents for organochlorinated compounds (such as organochlorinated pesticides, PCP, and PCBs). In a dynamic system, they are more useful than the traditional activated carbon. The presence of the amino group (primary or secondary) in the chemically modified chitosan is requisite to preserve the chitosan adsorptive properties for PCBs and PCP. The cross-linkage of chitosan through hydrophobic functional groups, such as the glutaraldehyde aliphatic chain, increases the adsorptive properties of the polymer. Nevertheless, this functional group must be sufficiently long so that the steric hindrance caused by cross-linkage does not impede adsorption. Through glutaraldehyde cross-linking and the

NaBH₃CN reduction of chitosan, the primary amino group is changed
to a secondary amino group and a hydrophobic functional group is introduced in the form of an aliphatic chain containing five CH_2-groups.
The PCB binding on the CHT-Glu-Red is the result of a combination of
strong ionic interaction and weak interaction of the hydrophobic type.
A PCP adsorption mechanism on the CHT-Glu-Red chitosan derivative
based on a weak dipole-dipole interaction is suggested. Moreover, the
nonwoven reactive filter K8WOT is a powerful and useful PCP and PCB
adsorbent. At the present time, it appears that the use of the cross-
linked chitosan-based filter is the most effective and the easiest way to
ensure PCB and PCP recovery from contaminated stream water.

REFERENCES

Asano, T. 1978. "Chitosan Applications in Wastewater Sludge Treatment," in
Proc. First Int. Conf. Chitin/Chitosan, R. A. A. Muzzarelli and E. R. Pariser,
eds., Cambridge, MA: MIT Sea Grant Report, 78-7, pp. 231–252.

Bough, W. A. 1975. "Production of Suspended Solids in Vegetable Canning
Waste Effluents by Coagulation with Chitosan," *J. Food Sci.*, 40:297.

Bough, W. A. 1976. "Chitosan, a Polymer from Seafood Waste for Use in Treatment of Food Processing Wastes and Activated Sludge," *Process. Biochem.*,
11(1):13.

Bough, W. A. and D. R. Landes. 1978. "Treatment of Food Processing Wastes
with Chitosans and Nutritional Evaluation of Coagulated By-Products," in
Proc. First Int. Conf. on Chitin/Chitosan, R. A. A. Muzzarelli and E. R.
Pariser, eds., Cambridge, MA: MIT Sea Grant Report, 78-7, pp. 218–230.

Bough, W. A., D. R. Landes, J. Miller, C. T. Young and T. R. Whorter. 1976. "Utilization of Chitosan for Recovery of Coagulated By-Products from Food Processing Wastes and Treatment Systems," *Proc. Natl. Symp. Food Process
Wastes 6th*, 22.

Clayton, J. R., Jr., S. P. Pavlou and N. T. Breitner. 1977. "Polychlorinated
Biphenyls in Coastal Marine Zooplankton: Bioaccumulation by Equilibrium
Partitioning," *Env. Sci. Technol.*, 11:676.

Davar, P. and J. P. Wightman. 1981. "Interaction of Pesticides with Chitosan,"
in *Adsorption from Aqueous Solutions*, P. H. Tewari, ed., New York: Plenum
Publishing Corporation, pp. 163–177.

Duinker, J. S., D. E. Schultz and G. Petrick. 1988. "Selection of Chlorinated
Biphenyl Congeners for Analysis in Environmental Samples," *Mar. Pollut.
Bull.*, 19:74.

Duursma, E. K. and M. Marchand. 1974. "Aspect of Organic Marine Pollution,"
Oceanogr. Mar. Biol. Ann. Rev., 12:315.

Giles, C. H., A. P. Da Silva and I. A. Easton. 1974. "A General Treatment and
Classification of the Solute Adsorption Isotherms, Part II: Experimental In-
terpretations," *J. Colloid Interface Sci.*, 47:766.

Hall, F., R. A. Downer and A. C. Chapple. 1992. "Evaluation of the Carbo-Flo® Waste Management System," *ASTM STP 112, Pesticide Formulations and Applications Systems,* Vol. 11, pp. 24–32.

Hiraizumi, Y., M. Takahashi and H. Nishimura. 1979. "Adsorption of Polychlorinated Biphenyls into Bed Sediment, Marine Plankton, and Other Adsorbing Agents," *Env. Sci. Technol.,* 13:580.

Hutzinger, O., S. Safe and V. Zitko. 1974. *The Chemistry of PCBs,* Boca Raton, FL: CRC Press.

Inoue, K., Y. Baba, K. Yoshizuha, H. Noguchi and M. Yoshizaki. 1988. "Selectivity Series by Crosslinking Copper(II)-Complexed Chitosan," *Chem. Lett.,* (1):1281.

Jeuniaux, Ch. 1963. *Chitine et Chitinolyse, un Chapitre de la Biologie Moléculaire.* Paris:Masson.

Keisuke, K., K. Yoshiyaki and I. Akihiko. 1986. "Studies on Chitin. IX. Crosslinking of Water-soluble Chitin and Evaluation of the Products as Adsorbents for Cupric Ion," *J. Appl. Polym. Sci.,* 31(5):1169.

Keisuke, K., C. Satoyuki, K. Yoshiyaki. 1988. "Studies on Chitin, Part 15, Improvement of Adsorption Capacity for Copper II Ion Br N-Nonanoylation of Chitosan," *Chem. Lett.,* (1):9.

Kemp, M. V. and J. P. Wightman. 1981. "Interaction of 2,4,-D and Dicamba with Chitin and Chitosan," *Va J. Science,* 32(2):34.

Knorr, D. 1983. "Dye Binding Properties of Chitin and Chitosan," *J. Food Sci.,* 48:36.

Koshijima, T., M. Yoneda, R. Tanaka and F. Yaku. 1973. "Chelating Polymers Derived from Cellulose and Chitin. II. Variation of the Amounts of Combined Metal Ions with Functional Group Densities of Cellulosic Chelating Polymers," *Cellul. Chem. Technol.,* 11(4):431.

Kurita, K., Y. Koejama and S. Chikaoka. 1988. "Studies on Chitin. XVI. Influence of Controlled Side Chain Introduction to Chitosan on the Adsorption of Copper(II) Ion," *Polym. J.,* 20(12):1083.

Lord, K. A. 1948. "The Sorption of DDT and its Analogues by Chitin," *Biochem. J.,* 43:72.

Masri, M. S. 1980. "Treatment of Hexavalent and Trivalent Chromium in Dye Bath," in *Proceedings of the 6th International Wool Textile Research Conference, Vol. 5,* Pretoria, S. A. pp. 487–499.

Masri, M. S. and V. G. Randall. 1978. "Chitosan and Chitosan Derivatives for Removal of Toxic Metallic Ions from Manufacturing Plant Waste Streams," in *Proc. First Int. Conf. on Chitin/Chitosan,* R. A. A. Muzzarelli and E. R. Pariser, eds., Cambridge, MA: MIT Sea Grant Report, 78-7, pp. 277–287.

McFarland, V. A. and J. U. Clarke. 1989. "Environmental Occurrence, Abundance and Potential Toxicity of Polychlorinated Biphenyl Congeners: Considerations for a Congener-Specific Analysis," *Environ. Health Perpect.,* 81:225.

McKay, G., H. S. Blair and A. J. Gardner. 1982. "Adsorption of Dyes on Chitin. I. Equilibrium Studies," *J. Appl. Polym. Sci.,* 27:3043.

McKay, G., H. Blair and A. Findon. 1986. "Kinetics of Copper Uptake on Chitosan," in *Chitin in Nature and Technology,* R. Muzzarelli, Ch. Jeuniaux and G. W. Gooday, eds., New York, London: Plenum Press, pp. 559–569.

Muzzarelli, R. A. A. 1973. *Natural Chelating Polymers,* Oxford: Pergamon Press.

Muzzarelli, R. A. A. 1977. *Chitin,* Oxford: Pergamon Press.

Muzzarelli, R. A. A. and F. Tanfani. 1981. "The Chelating Ability of Chitinous Materials from *Streptomyces, Mucor rouscii, Phycomyces blaheslecanus* and *Choanephora cacurbitarum," J. Appl. Biochem.,* 3(4):322.

Muzzarelli, R. A. A. and F. Tanfani. 1982. "The Chelating Ability of Chitinous Materials," *Proc. Second Intl. Conf. Chitin and Chitosan,* Sapporo, Japan, pp. 183–186.

Muzzarelli, R. A. A., F. Tanfani, M. Emanuelli and L. Bolognini. 1985. "Aspartate Glucan, Glycine Glucan and Serine Glucan for the Collection of Cobalt and Copper from Solutions and Brines," *Biotechnol. Bioeng.,* 27(8):1115.

Portier, R. J. 1986. "Chitin Immobilization Systems for Hazardous Waste Detoxification and Biodegradation," in *Immobilisation of Ions by Bio-Sorption,* Eccles and Hunt, eds., Chickertes: Ellis Horwood Ltd., pp. 229–243.

Portier, R. J. 1989. "Microorganisms for biodegrading toxic chemicals," August 22, 1989. U.S. Patent 4,859,594.

Portier, R. J. and K. Fujisaki. 1986. "Continuous Biodegradation and Detoxification of Chlorinated Phenols Using Immobilized Bacteria," *Toxicity Assessment,* 1:501.

Richards, A. G. and L. Cutkomp. 1946. "Correlation between the Possession of a Chitinous Cuticle and Sensitivity to DDT," *Biol. Bull. (Woods Hole, Mass.),* 90:97.

Safe, S. 1990. "Polychlorinated Biphenyls (PCBs), Dibenzo-*p*-dioxins (PCDDs), Dibenzofurans (PCDFs) and Related Compounds: Environmental and Mechanistic Considerations which Support the Development of Toxic Equivalency Factors (TEFs)," *CRC Crit. Rev. Toxicol.,* 20:133.

Sariaslani, F. S. 1991. "Microbial Cytochromes P-450 and Xenobiotic Metabolisation," *Adv. Appl. Microbiol.,* 36:439.

Seo, T. and T. Iijima. 1991. "Sorption Behavior of Chemically Modified Chitosan Gels," in *Biotechnology and Polymers,* C. B. Cebelein, ed., New York: Plenum Press, pp. 215–227.

Seo, T., H. Ohtake, T. Kanbara, K. Yonetake and T. Iijima. 1991. "Preparation of Permeability Properties of Chitosan Membranes Having Hydrophobic Groups," *Makromol. Chem.,* 192:2447.

Sparling, J. and S. Safe. 1990. "The Effects of *ortho*-chlorosubstituents on the Retention of PCB Isomers in Rat, Rabbit, Japanese Quail, Guinea Pig and Trout," *Toxicol. Lett.,* 7:23.

Thomé, J. P. and Y. Van Daele. 1986. "Adsorption of PCB on Chitosan and Application to Decontamination of Polluted Stream Waters," in *Chitin in Nature and Technology,* R. A. A. Muzzarelli, Ch. Jeuniaux and G. W. Gooday, eds. New York, London: Plenum Press, pp. 551–554.

Thomé, J. P., M. Honorez, J. L. Hugla and Y. Marneffe. 1991. "Bioaccumulation et Cinétique d'Adsorption des PCBs par une Espèce de Crustacé Planctonique d'Eau Douce: *Daphnia magna,*" in *3rd Conference Internationale de Limnologie d'Expression Française (C.I.L.E.F.), Hommage à F. A. Forel,* J. P. Vernet, ed., Morges (1991), pp. 171–174.

Thomé, J. P., J. L. Hugla and M. Weltrowski. 1992. "Affinity of Chitosan and Related Derivatives for PCBs," in *Advances in Chitin and Chitosan,* C. J. Brine, P. A. Sandford and J. P. Zikakis, eds. London and New York: Elsevier Applied Science, pp. 639–647.

Thomé, J. P., I. Thys, J. L. Hugla, J. Patry and M. Weltrowski. 1994. "Chemical Affinity of Chitosan Derivatives for PCBs in Natural Freshwaters: Comparison of the Sorption Efficiency on Chitosan and Bioconcentration by *Daphnia magna,*" in *Chitin Word,* Z. S. Karnicki, A. Wojtasj-Pajak, M. M. Brzeski and P. J. Bylakowski, eds. Bremerhaven, Wirtschaftsverlag, NW, pp. 255–265.

Thomé, J. P., J. Patry, I. Thys and M. Weltrowski. 1996. "New Chitosan Based PCB Adsorbents: A Synthesis," in *Advances in Chitin Science, Vol. 1,* A. Domard, Ch. Jeuniaux, R. Muzzarelli and G. Roberts, eds., Lyon-France: Jacques André Publisher, pp. 470–475.

Toussant, D. and M. Parent. 1986. "Le Traitement des Eaux Souterraines Contaminées par des Résidus de Pesticides," Centre de Documentation, Direction Environnement Hydro-Quebec, Quebec, Canada.

Tucker, E. S. and H. Saeger. 1975. "Activated Sludge Primary Biodegradation of PCB," *Bull. Environ. Contam. Toxicol.,* 14(6):705.

Uragami, T., F. Yoshida and M. Sugihara. 1983. "Studies of Synthesis and Permeabilities of Special Polymer Membranes. LI. Active Transport of Halogen Ions through Chitosan Membranes," *J. Appl. Polym. Sci.,* 28:1361.

Urano, K., E. Yamato, M. Tomegawa and K. Fujie. 1991. "Adsorption of Chlorinated Organic Compounds on Activated Carbon from Water," *Water Res.,* 26(12):1459.

Van Daele, Y. and J. P. Thomé. 1986. "Purification of PCB Contaminated Water by Chitosan. A Biological Test of Efficiency Using the Common Barbel, *Barbus barbus,*" *Bull. Environ. Contam. Toxicol.,* 37:858.

Waid, J. S. 1986. *PCB and the Environment,*" *Vol. 1,* Boca Raton, FL: CRC Press.

Weltrowski, M., R. Saint-André and J. Patry. 1993. "Etude sur les Méthodes de Traitement de l'Eau Contaminée," Saint-Hyacinthe, Canada: Textile Technology Centre, Internal Report, May 1993.

Weltrowski, M., J. Patry and R. Saint-André. 1994a. "Traitement des Eaux Contaminée au Pentachlorophénol," Saint-Hyacinthe, Canada: Textile Technology Centre, Internal Report, July 1994.

Weltrowski, M., R. Saint-André, J. Patry, R. Chénier and J. P. Thomé. 1994b. "Purification of Pentachlorophenol (PCP) Contaminated Water by Chitosan Cross-Linked in a Column or Filter System," in *Chitin Word,* Z. S. Karnicki,

A. Wojtasj-Pajak, M. M. Brzeski and P. J. Bylakowski, eds. Bremerhaven, Wirtschaftsverlag, NW, pp. 266–275.

Weltrowski, M., J. Patry and M. Bourget. 1996. "Reactive Filter for Textile Dyes Adsorption," in *Advances in Chitin Science, Vol. 1,* A. Domard, Ch. Jeuniaux, R. Muzzarelli and G. Roberts, eds., Lyon-France: Jacques André Publisher, pp. 462–469.

Yoshiyaki, K. and I. Akihiko. 1986. "Studies on Chitin. X. Homogeneous Cross-linking of Chitosan for Enhanced Cupric Ion Adsorption," *J. App. Polym. Sci.,* 31(6):1951.

Zikakis, J. P. 1984. *Chitin, Chitosan and Related Compounds,* New York: Academic Press.

Index

Affinity (*see also* microcapsules,
    PCP, PCB, plant protection)
  adsorption isotherms, 241–245
  adsorption studies, 238
  albumin for Cibacron Blue
    3GA dextran, 241
  concept of bioseparation using
    affinity capsule, 235
  preparation of affinity
    capsules, 236
  using microcapsules, 233–254
Agriculture (*see also* Food)
  food processing, 16–18
  inhibition of molting in
    insects, 141
  plant protection
    chitin and derivatives, 171
    antibacterial activity, 172
    antiphage activity, 180
    antiviroid activity, 178
    antivirus activity, 174
    stimulation of plant growth,
      183
  properties of chitin synthase, 155
    inhibitors, 156, 160
Antibacterial activity of chitin,
    172–174
  dependence on chitin derivatives,
    173

dependence on plant type, 175
Applications
  chitin, 32–47
  chitosan, 9–21

Cell encapsulation (*see also*
    microcapsules), 18
Chemical modifications
  N-phthaloyl-chitosan, 108
  soluble precursors, 103
  tosyl- and iodo-chitins, 106
  water-soluble chitin, 104
Chitin (*see also* structure,
    applications, wastewater)
  chitin-protein interaction,
    271–273
  β-chitin, 80–86
  maintenance of structure during
    purification, 260–269
  optical confirmation, 269–271
  properties, 57–73
    activity of chitin-specific
      enzymes, 60–72
    reactions of β-chitin, 80–86
    reactions of chitosan derived
      from β-chitin, 84–86
  structure and isolation, 255–275
  tosyl- and iodo-chitins, 106
  water soluble chitin, 104

Chitinase inhibitors, 190–193
  Lucilia cuprina, 187–190
  veterinary importance, 185
Chitinolytic enzymes, 185
  inhibitors, 190
  Lucilia cuprina chitinase, 187
  molecular weight, 193
Chitosan derivatives, 89–100
  antibacterial, antiviruses,
    antiviroid, antiphage
    activity, 172–182
  N-phthaloyl-chitosan, 108–110
Chitosan membranes (*see also*
    microcapsules), 18–21
Chironomus tentans, 160
  biosynthesis, 160
Commercial products, 46–47
  absorbable suture, 42
  additive for food, 43
  agricultural products, 38
  current commercial uses of
    chitin, chitosan, 33
  wound dressing, 41
  wound healing, 36
Copolymerization
  graft copolymers, 297–305
  of aniline with chitosan, 303
  of *N*-carboxy anhydrides (NCAs)
    with chitin, 300
  of vinyl monomers with chitin,
    298
  of vinyl monomers with
    chitosan, 302

Dietary food additive (*see also*
    agriculture)
  hypocholesterolemic effect,
    115–120
  hypouricemic effect of chitin and
    chitosan, 123
  treatment of celiac disease, 120
Drug carriers
  6-*O*-Carboxymethyl-chitin-
    mitomycin
    conjugate, 223
  drug release, 217
  chitosan microsphere
    biodegradation, 209
    preparation and properties,
      206–209
  N-Succinyl-chitosan-mitomycin
    C conjugate, 223

preparation of chitosan-
    cytarabine conjugate,
    212–217

Endochitimase, 60
Environmentally friendly
    application, 31–49

Filaments (*see also* polymers,
    textiles)
  prepared from chitin derivatives,
    281–295
  density and optical properties,
    290
  electrical properties, 290
  macrostructure, 285
  microstructure, 285
  physical properties, 288
  physicochemical properties,
    293
  thermal properties, 290
Food (*see also* agriculture, dietary)
  chitosan as a food additive, 115
  hypocholesterolemic effect, 120
  hypouricemic effect, 123
  enhanced food production
    phytohormones, 131
  seed coating, 129

Industrial pollutants, 309
Inhibition of alfalfa mosaic virus,
    176–177
Insect chitin synthase, 155–167
  biological activity of inhibitions,
    156
  effect of inhibitions, 160
Insects (*see also* agriculture)
  inhibition of molting, 141–152
  molt-inhibited *Manduca larvae*,
    144
Isolation (*see also* affinity)
  albumin from bovine plasma,
    250

Medicine (*see also* drug carriers),
    236, 238
  affinity microcapsules, 205
  chitosan alginate capsules, 236
  chitosan as drug carriers, 206,
    212
  chitosan-drug conjugate, 206,
    212

isolation of bovine serum
albumin, 238
Microcapsules, microbeads (*see
also* affinity), 206, 212,
236–238
equilibrium behaviour in
adsorption studies, 243
viable cell immobilization, 20
Molecular weight
chitinase, 193–199

PCB (polychlorinated biphenyls)
(*see also* wastewater)
preliminary studies with chitin
and chitosan, 313
sorption capacity on chitosan,
317–322
chemically modified chitosan,
314
PCP (pentachlorophenol) (*see also*
wastewater, pollutants)
adsorption on chitosan
derivatives, 322–326
sorption capacity, 324
Pharmaceutics, 13
Physicochemical properties of
chitosan, 6
Plant antiviroid activity of
chitosan, 178–179
Plant antivirus activity of
chitosan, 174–175
Plant growth (*see also* agriculture
and plant protection)
biostimulation by chitosan,
183
Plant protection (*see also*
agriculture), 171–182
chitin and derivatives
antibacterial activity, 172
antiphage activity, 180
antiviroid activity, 178
antivirus activity, 174
stimulation of plant growth,
183
Pollutants
organochlorine, 311–313
Polymers, 297
filaments of chitin derivatives,
281–295
graft copolymerization on chitin,
298

graft copolymerization of
chitosan, 302
Properties (*see also* structure,
chitin, chitosan,
agriculture)
chitin
activity of chitin specific
enzymes, 60–72
chitin synthase, 155
reactions of β-chitin, 80–86
chitin synthase, 155
inhibitors, 156, 160
chitosan, 5–9, 31–32
reactions, 84–86
chitosan derivatives, 89–100
antibacterial, antivirus,
antiviroid, antiphage,
172–182
N-phthaloyl-chitosan, 100–108
molecular weight of chitinolytic
enzymes, 185

Reactivity characteristics (*see also*
chitin, chitosan)
β-chitin, 79–86
acetolysis, 84
deacetylation, 80
full acetylation, 81
isolation, 80
N-acetylation, 80
tosylation, 82
tritylation, 83

Seed-coating using chitosan
California rice, 134
enhanced food production,
134–137
Louisiana rice, 129–137
phytohormones, 135
Separation of chitinases, 63
Solution properties of chitosan,
89–100
O,N-carboxymethylchitosan, 93
Stimulation of plant growth, 183
Structure (*see also* chitin,
chitosan)
chitin, 255
chemical identity, 255
chitin-protein interaction, 271
optical conformation, 269
purification, 260
chitin-specific enzymes, 57–72

Textiles, 281
  chitin filaments, 292

Veterinary applications, 185–198

Wastewater, 309
  chlorine content, 317

organochlorinated compounds,
  310
PCB, 311, 313
PCP, 317, 320
sorption of PCB on chitin and
  chitosan, 313, 314, 317